建筑科学基础

1

2

3

4

5

6

7

8

9

10

11

12

13

14

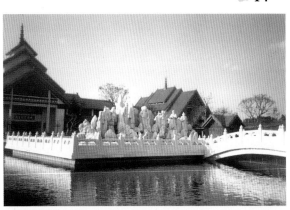
15

16

17

18

19

20

21

22

23

24

25

26

27

28

29

30

31

32

33

34

35

36

37

39

38

40

41

42

43

44

图45

图46

图47

图48

📷 49

📷 50

📷 51

📷 52

📷 53

📷 54

关于彩图的说明

图是环境艺术、城市规划、建筑设计、建筑装饰等专业的第一语言,尤其传递立体和彩色信息时,文字绝难达到彩图的效果。

彩图:No.1. 依据生态可持续建筑原理散水种植的效果(传统观念散水是从不种植的),你一看就会有"百闻不如一见"的效果。

No.2. 利用月光加强夜艺术表演效果。

No.3. 中国石窟洞建筑文脉。

No.4. 美国一数据中心:世界闻名的掩土绿化太阳房.

No.5. 庭园设计:小桥、静水、楼家。

No.6. 四环图案花园。
No.7. 清华大学改荒岛为荷池、廊、阁。

No.8. 北大变荒角为美丽校园一角。

No.9. 长廊艳花丛无限美享受——杭州机场出入口。

No.10. 深圳某广场绿化。
No.11. 云南丽江—古街。

No.12. 丽江黑龙潭——世界闻名景点。

A General Explain on Color Pictures

Pictures are the first language of the professions of Environmental Art, Urban Planning, Building Design and Building Decoration, etc. especially in transmitting sterreo and color informations, words however very hardly get the effect as color picture does.

Color pictures: No.1. An apron planting according to eco-sustainable architecture principle (in traditional concept, apron has never been planted), as you have a look you'll fell "it's better to see once than hear a hundred times".

No.2. Using moon light to increase the effect of night performing arts.

No.3. Chinese stone cave architectural context.

No.4 An America Data Center: A world famous earth sheltered solar greening building.

No.5. Garden design: small bridge, static water and buildings.

No.6. Four-round-pattern garden.

No.7. Tsinghua Univ reconstructed the wasteland into lotus pool, corridor and pavilion.

No.8. Peking Univ made the waste corner into a beautiful campus corner.

No.9. A long corridor clustered flowers give you a limitless enjoyment of beauty—the inlet and outlet of Hangzhou airport.

No.10. A square greening in Shenzhen.

No.11. An ancient street in Lijiang, Yunnan.

No.12. Lijiang Heilong Pool——a world famous landscape.

No.13. 重檐牌楼傍巨石，巍然屹立。

No.14. 好精美的石雕牌楼！

No.15. 黄琉璃顶、红粉墙、白玉拱，对比强烈而又谐和。

No.16. 琉璃叠顶、白玉群雕、水镜倒影——环艺设计良例。

No.17. 西安人民剧院西墙绿化。夏遮阳，冬日照，有利节能。

No.18. 回廊漫步水池边，随时可小歇。

No.19. 西安新城广场，市民最大活动中心。

No.20. 好美的姑娘！鲜艳的民族服饰，流利的英语，导游的精英。

No.21～23.（丽江），No.24.（北京）.环境艺术设计选景。

No.25, No.27. 不同民族风格庭院设计。

No.26, No.28. 匈牙利墙面、窗户装饰。

No.29, No.30. 作者的掩土绿化实验房。绿色面积不减少反而增多。

No.31～32. 昆明花市，全国最大供花基地。

No.33～34.（济南），No.35（布达佩斯）。庭院设计选景。

No.36. 峨嵋山，顶峰俯瞰云海间，波涛奇景无限。

No. 13. The pailou with the piled eaves against the tall rock, they stand loftily amd firm.

No. 14. How fine the stone carving pailou!

No. 15. Yellow glazed tile roof, red renderd wall, white marble arch, they compare strongly with each other but show harmony.

No. 16. Glazed tile roofs piled, white marble grouped carvings (statues), water-mirror inverting images. A good example of environmental art design.

No. 17. Xi'an People Theatre west-facing wall greening, summer shading, winter sun shine that benefits energy saving.

No. 18. Strolling on the winding walkway by the pool, anytime may have a short rest.

No. 19. Xi'an Xincheng Square, the biggest action center for the citizen.

No. 20. How beautiful the girl! Bright colored national dress, fluent English, a pick of the guiders.

No. 21～23. (Lijiang), No. 24 (Beijing). Selective sights of environmental art design.

No. 25, No. 27. Garden designs of different national style.

No. 26, No. 28. Hungary wall and window decoration.

No. 29, No. 30. The writer's earth sheltered building planted, green area didn't decrease but increase.

No. 31～32. Kunming flower market, the biggest flower-supply base in China.

No. 33～34 (Jinan), No. 35 (Budapest). Seletive sights of garden design.

No. 36. Emei Mountain, topping looking down, swing eyes over the cloud-sea, wonderful wave sights endless.

No.37~38. 西安钟鼓楼广场及地下购物中心。

No.39. 小桥流水鲜花丛,静街沿岸翠柳行(丽江)。

No.40. 墙面爬蔷薇,丛花满地生。

No.41. (西安). No.42(北京),两环艺设计例。

No.43. 广州白云宾馆室内花园,一个极好的室内设计。

No.44. 云南石林剑峰——世界闻名景点之一。

No.46. 飞檐翘指苍蓝天,重砖叠砌拱门关。

No.45. 鲜艳的民族服饰小演员的姿态,闪亮在深绿的背景前。

No.47. 朝阳下,美服装,第一时间观红场。

No.48. 没有老师、没有阿姨,服饰乱七八糟,站立歪歪扭扭,但却天真、澜漫、可爱!

No.49. No.50, No.53(西安), No.51, No.52(济南). 两城护城河环境视景选:美、静、和平。

No.54. 窗台四盘鲜花,美感立增。

※　※　※　※　※

以上彩图乃作者在力所能及的条件下收集的体现设计理念和原理的实例,它们不是一一对应某一细节的图示,故特综合作此《一般说明》。

No. 37 ~ 38. Xi'an Tower-Drum Building Square and the underground shopping center.

No. 39. Small bridge, water flowing, flower thickly growing and green willows lined the bank of the quiet street(Lijiang).

No. 40. Rose climbing the wall, flowers growing fully on the ground.

No.41(Xi'an), No. 42(Beijing). Two examples of environmental art design.

No. 43. Indoor garden, Beiyun Hotel, Guangzhou, an excellent indoor design.

No. 44. Yunnam Stone-forest Sword-top, one of the world famous sights.

No. 46. Upturned eaves pointing towards blue sky, heavy bricks laid the arch gate.

No. 45. Bright colored national dress, little player style, flashing at the front of the deep green background.

No. 47. Under morning sun shining, in beautifull dress, at the first time to visit the Red Square.

No. 48. No teacher, no auntie, dressed at sixes and sevens, standing in shapeless and twisted shapes. But they look innocent, artless and lovely!

No. 49. No. 50, No. 53(Xi'an), No. 51, No.52(Jinan). Selections of the moat surround of the two cities:beauty, quiet and peace.

No. 54. Four plate flowers on windowsill immediately increase beauty.

※　※　※　※　※

The above color pictures are collected by the writer within his capability, they are practical examples embodying the design idea and principle, they are not any figures corresponding to any details one by one, so we comprehensively write this《General Explain》.

汉英双语教材
BILINGUALISM OF CHINESE ENGLISH BOOK

建筑科学基础
FOUNDATIONS OF ARCHITECTURAL SCIENCE

夏 云　夏 葵　编著

中国建材工业出版社

图书在版编目(CIP)数据

建筑科学基础/夏云,夏葵编著. —北京:中国建材工业出版社,2005.4(2016.8 重印)
ISBN 978-7-80159-820-2

Ⅰ.建… Ⅱ.①夏…②夏… Ⅲ.建筑学-汉、英 Ⅳ.TU

中国版本图书馆 CIP 数据核字(2005)第 031283 号

内 容 摘 要

本书主要内容:1.环境·人·建筑。2.建筑科学基础:地基、基础;墙:一般构造,防水防潮,散水绿化;冬季建筑热学;隔墙·隔空气声·噪声控制;楼板·隔撞击声;阳台、雨棚及其绿化效益;楼梯·建筑防火;屋顶·夏季防热;门窗·建筑采光·窗温室生态效益;建筑变形与抗变形;表面处理;厅堂音质。

本书适用于环境艺术、城市规划、建筑学以及建筑装饰等专业学生汉——英双语教材(已在西安建筑科技大学开课 8 年,1997—2004)。

AN ABSTRACT OF THE CONTENTS

Main contents of this book: 1. Environments·Being·Architecture. 2. Foundations of Architectural Science: Ground Bases, Foundations; Walls: General Construction, Protection of Wall from Water and Moisture, Apron Greening; Building Heat in Winter; Partitions·Insulation of Airborne Sound·Noise Control; Floors·Insulation of Impact Sound; Balcony, Canopy and Their Greening Benefits; Stairs·Building Fireproof; Roof·Summer Heat Insulation; Doors/Windows·Building Daylighting; Window Greenhouse Eco-benefits; Deformation and Anti-deformation of Buildings; Surface Treatments; Acoustic Properties of Halls.

This book is suitable to be the teaching book for the students of Environmental Art, Urban Planning, Architecture and Building decoration, etc. (This book has been used as teaching book for 8 years, 1997-2004, in the Xi'an University of Architecture and Technology).

建筑科学基础

夏云 夏葵 编著

出版发行:中国建材工业出版社
地　　址:北京市海淀区三里河路 1 号
邮　　编:100044
经　　销:全国各地新华书店
印　　刷:北京鑫正大印刷有限公司
开　　本:787mm×1092mm　1/16
印　　张:10.25　插页:4
字　　数:246 千字
版　　次:2005 年 7 月第 1 版
印　　次:2016 年 8 月第 3 次
定　　价:35.00 元

本社网址:www.jccbs.com.cn
本书如出现印装质量问题,由我社发行部负责调换。联系电话:(010)88386906

夏云简介

Profile of Xia Yun, one of the authors

夏云教授,1954~2004 年教学 50 年。1950~1954 年在北大、清华学习,1954 年毕业于清华大学建筑学专业,留校任教至 1957 年。自 1957 年开始在西安建筑科技大学任教,现兼任建筑学院教学督导。

教　学:汉英双语讲授:生态可持续建筑;建筑科学基础;建筑构造;建筑物理;建筑防火;专业英语。

访问讲学:国内 14 所院校;国外 3 所院校。

出　书:节能节地建筑基础;节能节地建筑(英语);生态可持续建筑及建筑科学基础(汉-英双语);英译汉两册:掩土建筑;建筑科学基础。

辅助教材:幻灯教材 5 集;建筑声学试验(英译汉 35 万汉字)。

科　研:自 1980 年,论文 40 余篇(率研究生在 12 次相关国际会议上宣读 28 篇)。

Professor Xia Yun, 1954~2004, teaching 50 years. 1950~1954 learning in Beijing Univ & Tsinghua Univ, Architecture Diploma, 1954, Tsinghua Univ, and teaching there till 1957. Since 1957 teaching in Xi'an Univ of Arch & Tech. Now a part-time educational inspector of the Architecture School.

Teaching: Bilingualism of Chinese-English teaching:Eco-sustainable Architecture;Foundations of Architectural Science; Building Construction; Building Physics; Building Fireproof; Professional English.

Visiting lectures: Inland 14 times to 14 colleges & universities; abroad 3 times to 3 ones.

Publications: Books: Foundations of Energy & Land Saving Buildings; Energy & Land Saving Buildings (English); Eco-sustainable Architecture and Foundations of Architectural Science (bilingualism of Chinese-English); 2 Translations English to Chinese: Earth Sheltered Buildings; Foundations of Architectural Science.

Assistants: 5 collections of teaching slides; Building Acoustic Experiment (translation English to Chinese 350 thousand Chinese words).

Researches: Since 1980, more than 40 papers (leading postgraduates read out 28 papers, 12 times on relative international conferences/congresses).

专　　利：与女儿夏葵发明《太阳能水土热惯性富氧自然空调构造》获国家发明专利。

荣　　誉：部级科技进步二等奖一项；与他人合作获联合国教科文组织与国际建协主办的《为可持续未来社区进行策划》国际设计竞赛专业组第3名；国务院突出贡献特殊津贴；被美国传略所（abi）和英国国际传略中心（IBC）选为"20世纪突出人物"之一；2002年，（IBC）授予"终身成就"奖。

Patent: With daughter Xia Kui invented "An Oxygen-rich Natural Air Conditioning Construction by Solar Energy & the Thermal Inertia of Water & Earth" has won the Chinese Patent.

Honours: A 2nd prize of Advanced Science and Tech of Ministry Level; With others won the 3rd professional prize of UNESCO & UIA, International Design Competition on "Design for the Sustainable Future Settlement"; Special Subsidy of the State Council of China for Outstanding Contribution; One of the "Outstanding People in the 20th Century" by USA ABI and UK IBC; 2002, "Lifetime Achivement Award" by IBC.

前 言
PREFACE

当代科技发展的与时俱进,必然引起传统教材新陈代谢的改革。本书——《建筑科学基础》就是在这种形势下产生的,它是对传统教材《建筑构造(含相关材料)》、《建筑物理》、《建筑防火》精简、更新、综合而成。

本书特点有三:

1. 消除重复

如房屋的保温、隔热、隔声、采光、遮阳等在《建筑构造》与《建筑物理》中均有重复,但在《建筑构造》中缺乏系统理论,在《建筑物理》中又缺乏具体构造;又如楼梯,在《建筑构造》中什么是"可封闭楼梯"、"防烟楼梯"只字不提,而在《建筑防火》中又缺乏具体构造……余例从略。

本书对此进行了精简、综合,消除了重复。

2. 精简、更新

上述传统三书共 200 余万字,作者讲授期,共 200 学时,平均每学时学生要阅读万余字,可见内容之多!

Today, catching time the science-technology has been developing and certainly causing the metabolic reform of the conventional teaching books. This book 《Foundations of Architectural Science》(FAS) has been arisen under this situation. This FAS is composed by the conventional teaching books: 《Building Construction (including relative materials)》(BC), 《Building physics》(BP) and 《Building Fireproof》(BF). Their original contents have been simplified, renewed and combined.

This book (FAS) has three specialities:

1. Repeats cancelled

Such as building winter and summer insulation, sound insulation, daylighting, summer shading, etc. in both BC and BP have existed repeats, but weak principle in BC and weak concrete construction in BP; about stairs, what means "closable staircase?" "smokeproof staircase"? in BC no any word, in BF no any concrete construction… other examples omitted.

The FAS has cancelled the above repeats by the way of simplifying and combining.

2. Simplification and renewal

The above three conventional books altogether have more than 2.0 million words, during my teaching the total teaching hours were 200 hours, it meant that average each hour each student must read more than 10

本书除对上述传统教科书中过多内容进行精简外并对某些过时的理论、构造进行了更新,如散水、外墙、窗户、阳台、遮阳、屋面等建筑构件,传统理论重点讲它们的围护和某些使用功能,但却忽略了它们通过反射、排水,浪费了大量太阳能与雨水(含能物流)(作者曾估算,全国建筑按中等水平,一年反射浪费的太阳能如果转为电能就够全国2亿5千万户,每户年用电2000 kWh,用12年!),因此,按照生态可持续原理,这些构件都不是可持续构件,必须改革,彩图1为一例,将传统散水加以种植,不仅利用了太阳能和雨水,还有生态正效应,这种散水就从非可持续构件变成了可持续构件。

3. 汉—英双语

双语教学是保持教学高水平的关键之一。10余年实践证明该课深受学生喜爱(虽然他们英语水平不同)。两点建议:(1)学生英语水平较低也可上双语课,教师可采用可行的启发、讨论使学生各自得到专业知识和英语水平的提高;(2)年轻教师英语基础不深厚要敢上。勤奋备课,师生互助,在教学中提高。果如此,君必成功常伴!

thousand words, very clearly the contents are too much!

In the FAS the overmuch contents in the above conventional teaching books have been simplified, besides, some outdated theories and constructions also have been renewed. For an example, on building members of apron, exterior wall, window, balcony, shade and roofing, etc. the conventional theories only focus on their functions of envelope and some uses, however their much wastes of solar energy and rainwater (mass flow contained energy) by reflection and drainage have still been forgotten (we have estimated that at a mean level, the all buildings of China, each year reflected the solar energy which if could be converted into electricity that might enough supply all Chinese 0.25 billion families' electricity-use, per family per annum 2000 kWh for 12 years!). So according to eco-sustainable principles, these building members are not sustainable members, they should be reformed. No.1 color picture is an example, the apron has been planted which not only can use solar energy and rainwater but also has positive eco-effect. This apron has become from non-sustainable member into sustainable one.

3. Bilingualism of Chinese-English

Bilingual teaching is one of the key links for keeping high level teaching. The practice of more than 10 years has proved that students very like the course (although their English levels are different). Two suggestions: (1) To the students of rather low English level, the bilingual course can be also taught, the teacher may use available enlightening and discussing ways to help every student to gain respective increases of professional knowledge and English level; (2) Young teacher should

感谢西安建筑科技大学建筑学院与艺术学院多年的大力支持;感谢同学们的好建议;感谢该校印刷厂多年出版本课内部教材;感谢中国建材工业出版社的大力支持,正式公开出版此书;特别感谢我的家人,承担全部家务并帮助双语书写。

热诚欢迎指缺点,提建议。

作　者
2004年10月28日

be brave teaching the course although your English basis may be not very deep. Proper course with diligence, teachers and students help each other, advance you in teaching, if done so, successes certainly always follow you!

Thanks to the College of Architecture and College of Art, Xi'an University of Architecture and Technology for many years' great support; thanks to the students' good suggestions; thanks to press of the university for many years publishing the inner teaching books; thanks to the China Building Materials Industry Press for the great supporting to publish the book in due form and openness. Especially thank my family for bearing the whole house work and helping the bilingual writing.

Very warmly welcome to point shortcomings and provide suggestions.

The authors
28/10/04

目 录
CONTENTS

1 环境·人·建筑 ……………………… 1
  1.1 环境 ………………………………… 1
  1.2 围绕人类的自然环境 ……………… 2
    1.2.1 天 …………………………… 2
    1.2.2 地球 ………………………… 5
  1.3 房屋及其环境是如何
      建起来的? ………………………… 15
    1.3.1 设计分类…………………… 15
    1.3.2 施工的主要任务 …………… 17
    1.3.3 简言之……………………… 17
    1.3.4 如何感知房屋
          及其环境? ………………… 18
  1.4 建筑物分类,分等及
      防火等级…………………………… 18
    1.4.1 建筑物分类………………… 18
    1.4.2 建筑物分等………………… 20
    1.4.3 建筑物防火分级…………… 21
    1.4.4 建筑物组成构件…………… 22
    1.4.5 建筑物平、立、剖面
          表示法……………………… 24
2 建筑科学基础……………………… 25
  2.1 地基 ………………………………… 25
  2.2 基础 ………………………………… 27
    2.2.1 基础的类型………………… 27
    2.2.2 基础最小埋置深度………… 29

1 ENVIRONMENTS·BEING·
   ARCHITECTURE ……………… 1
  1.1 Environments ……………………… 1
  1.2 Natural Environments around
      Human Race ……………………… 2
    1.2.1 Sky ………………………… 2
    1.2.2 Earth ……………………… 5
  1.3 How to Build a House and Its
      Environments? …………………… 15
    1.3.1 Design classification ……… 15
    1.3.2 Construction main jobs …… 17
    1.3.3 In brief …………………… 17
    1.3.4 How to experience houses and
          their environments? ……… 18
  1.4 Types, Ratings and Fireproof Classes
      of Buildings ……………………… 18
    1.4.1 Types of buildings ………… 18
    1.4.2 Ratings of buildings ……… 20
    1.4.3 Fireproof classes of
          buildings ………………… 21
    1.4.4 Components of a
          building …………………… 22
    1.4.5 Showing of a building plan,
          facade and section ………… 24
2 FOUNDATIONS OF ARCHITECTURAL
   SCIENCE …………………………… 25
  2.1 Ground Bases ……………………… 25
  2.2 Foundations ……………………… 27
    2.2.1 Foundation types ………… 27
    2.2.2 Minimum buried depth of
          foundation ………………… 29

2.3	墙 …………………………… 32	
	2.3.1 黏土砖墙(简称砖墙)……… 32	
	2.3.2 墙身防水防潮………… 36	
2.4	冬季建筑热学……………… 40	
2.5	隔墙·隔空气声……………… 54	
2.6	噪声控制…………………… 61	
2.7	楼板·隔撞击声……………… 70	
	2.7.1 一般讨论………………… 70	
	2.7.2 钢筋混凝土楼板………… 71	
	2.7.3 钢筋混凝土楼板类型与施工……………………… 73	
	2.7.4 预应力钢筋混凝土楼板…… 77	
	2.7.5 撞击声隔减……………… 79	
	2.7.6 阳台、雨棚及其绿化效益………………………… 79	
2.8	楼梯………………………… 80	
2.9	建筑防火…………………… 85	
2.10	屋顶………………………… 95	
2.11	建筑夏季防热……………… 104	
2.12	门窗………………………… 109	
2.13	天然采光…………………… 115	
2.14	窗温室生态效益…………… 121	
2.15	建筑物变形与抗变形……… 121	
	2.15.1 地震…………………… 122	
	2.15.2 不均匀沉降…………… 125	
	2.15.3 热变形………………… 125	
2.16	表面处理…………………… 127	
	2.16.1 万事万物均需表面处理…………………… 127	

2.3	Walls …………………………… 32	
	2.3.1 Clay brick walls (brick walls) …………… 32	
	2.3.2 Waterproofing and dampproofing of walls …………… 36	
2.4	Building Heat in Winter ……… 40	
2.5	Partition Walls·Insulation of Airborn Sound ………………………… 54	
2.6	Noise Control ………………… 61	
2.7	Floors·Insulation of Impact Sound ………………………… 70	
	2.7.1 General discussion ……… 70	
	2.7.2 RC floors ……………… 71	
	2.7.3 RC floor types and construction …………… 73	
	2.7.4 Prestressed RC floors …… 77	
	2.7.5 Insulation of impact sound ………………… 79	
	2.7.6 Balcony, canopy(weather shed) and their greening benefits … 79	
2.8	Stairs ………………………… 80	
2.9	Building Fireproofing ………… 85	
2.10	Roofs ………………………… 95	
2.11	Summer Heat Insulation of Buildings …………………… 104	
2.12	Doors and Windows ………… 109	
2.13	Daylighting(Natural Lighting) ……………… 115	
2.14	Eco-Benefit of Window Greenhouse ………………… 121	
2.15	Deformation and Anti-deformation of Buildings ………………… 121	
	2.15.1 Earthquakes …………… 122	
	2.15.2 Unequal settlement …… 125	
	2.15.3 Thermal deformation … 125	
2.16	Surface Treatments ………… 127	
	2.16.1 All things have to take surface treated …………… 127	

2.16.2	建筑及其环境表面处理应考虑的因素 …… 127	2.16.2	Factors should be considered for the surface treatments of buildings and environments …… 127
2.16.3	表面处理法 …… 128	2.16.3	Surface treatment methods …… 128
2.17	厅堂音质 …… 136	2.17	Acoustical Properties of Halls …… 136
2.17.1	室内声音的表演 …… 136	2.17.1	Sound playing in a room …… 136
2.17.2	厅堂音质 …… 137	2.17.2	Acoustical properties of halls …… 137
2.17.3	著名观演厅声学数据 …… 138	2.17.3	Acoustical data of some well known halls …… 138
2.17.4	吸声 …… 143	2.17.4	Sound absorption …… 143

1 环境·人·建筑

1 ENVIRONMENTS·BEING·ARCHITECTURE

1.1 环境

"环境"一词的含义很广,根据不同领域的各自目的有多种多样的划分,如:

物质环境
精神环境
政治环境
经济环境
自然环境
人为环境
生态环境等

一切科学与艺术或美学工作都应"以人为本",注意！不是"以人为霸"。

从"以人为本"的观点来看人聚环境乃环境的中心,而建筑乃人聚环境的主体。因此,环境和建筑设计者以及规划师们都必须对建筑及其环境要有相应的了解,对相关建筑科技要有必要的掌握。

1.1 Environments

The meaning of the word "Environment" is very broad.

According to respective aims of different fields, environments have been variously divided such as:

Physical environment
Spirit environment
Political environment
Economic environment
Natural environment
Built(man-made)environment
Ecologic environment,etc.

All works of science, art and aesthetics should take "serving people as the root duty", heed! not "take being as the tyrant".

In the view of "serving people as the root duty", people living environment is the centre of environments, buildings are main bodies of the people living environment. So the environment and building designers and planners must have a suitable understanding of architecture and its environment, and master the needful architectural science – technic knowledge.

1.2 围绕人类的自然环境

1.2.1 天

(1) 太阳

太阳质量为 2.2×10^{27} t,约为地球质量的 33 万倍,占太阳系总质量的 99.86%,最主要的成分是氢占 78%,氦占 20%。

太阳直径约为 139 万 km,是地球直径的约 109 倍,太阳体积约为地球体积的 130 万倍,太阳至地球平均距离为 14,960 万 km。太阳表面温度 6000℃,核心温度据推测为 1000~2000 万 K。太阳能由其内部热核聚变过程产生,在该过程中,氢转变为氦,每秒有 400 万 t 物质转变为约 36×10^{22} kW 能量(爱因斯坦质能关系 $E = mC^2$,m 为质量 kg,$C = 3 \times 10^5$ km/s),如此惊人巨大的能量向太空辐射,其中约 22 亿分之一到达地球大气层外表面。穿过大气层抵达地面的能量约为 8.5×10^{13} kW,这就是地球生物及矿物的最原始的能源。

太阳能是最伟大的、干净、安全、永久性可持续能源。

没有太阳就没有万紫千红的世界。

(2) 月亮

月球是地球的天然卫星,其表面重力仅为地球表面重力的 1/6,质量是地球的 1/81。月球上无空气,无绿色植物,月面向阳(白天)

1.2 Natural Environments around Human Race

1.2.1 Sky

(1) Sun

Solar mass is 2.2×10^{27} t, about 0.33 million times that of the earth, occupies 99.86% of the total mass of the solar system. Hydrogen and helium are by far the most abundant, representing over 78 and 20 percent of the solar mass, respectively.

Solar diameter is about 1.39×10^6 km, almost 109 times that of the earth. Sun to earth mean distance is 0.1496 billion km. Solar surface temperature is about 6000℃, its core temperature is thought to be 10~20 million K.

Solar energy is generated within the sun in a thermonuclear fusion process, hydrogen transforms into helium, every second approximately 4 million t materials convert to energy of some 36×10^{22} kW (Einstein mass - energy conversion law: $E = mC^2$, m: kg, $C = 3 \times 10^5$ km/s). Within so whacking energy radiating to space, about one over 2.2 billion of the energy reaches earth atmosphere's outer surface. Through the atmosphere, 8.5×10^{13} kW energy reaches ground, this energy is just the original energy of earth living things and mines.

Solar energy is the greatest clean, safe and forever sustainable energy!

No sun no the world with a riot color!

(2) Moon

Moon is a natural satellite of earth. The gravity on moonscape is only one sixth of that on earth ground, moon's mass is only one

时温度高达 130~150℃,晚上低到负 180℃。最近发现月球两极可能有水冰存在。

移民月球是 21 世纪地球人类计划的重大太空行为之一。美、法工程人员正合作在月球就地取材建月球建筑。一美籍华裔工程师利用美国宇航员从月球带回的尘土已试制出月球水泥样品(从地球运 1t(吨)材料到月球要 5000 万美金)。

月球是宇宙科学研究的理想基地。月球上无空气,形成不了风,故月球表面虽有很多月球尘埃,也不会飞到空间降低能见度。在月球上,天文望远镜的分辨能力可以达到原"哈勃"太空望远镜的 10 万倍以上。

月球还是一个矿物较丰富的资源地,有 60 多种矿物。

上月球搞科研,如同去南、北极搞科研一样,我们决不放弃。

月球对地球主要有下列影响：

1)维持地球正常运行
太阳系中,天体的自转与公转运行都处在相互引力场作用的动态平衡中。

设若没有月亮,地球将发生滚翻运行,四季乱序,难以生存。

over eighty-one of that of the earth. On moon, no air, no green plants. When facing sun, moonscape temperature may reach 130~150℃, at night, may drop down to negative 180℃, in the two poles of the moon, recently has been found there might exist ice of water.

Immigrating people onto the moon is one of the great actions of earth man in the 21st century. The engineers of America and France have been cooperatively researching to build moon buildings with local materials on moon. An American engineer of Chinese origin has successfully got the cement sample made of the moon soil brought back by the American spaceman(to carry one t material to moon will cost USD 50 million).

The moon is an ideal site for space science researches. On moon no air so no wind, although there much dust on moonscape there's no any dust flying up to reduce the visibility. On moon the distinguishability of space telescope is more than 0.1 million times that of the original Hubble Space Telescope.

Moon is also an abudant resource of mines, there are more than 60 kinds of minerals.

Going to moon to do research as that going to the Antarctic Region and Arctic Region to do research we never give up.

Moon has the following main effects on earth:

1) keeping earth normally moving

In solar system, the celestial bodies revolve round the sun and on their own axes, they are kept in dynamic equilibrium by their gravitational field acted one another.

If there were no moon, the earth would move in roll and toss, and the moving rule of the four seasons would be confused, all living

2)潮汐作用

月亮的引力作用使地球海水不能与地球自转方向同步运行，而是相对于海底逆向而行，形成潮汐现象。

利用潮汐现象蓄水发电是一项可再生的可持续能源。

3)地震影响

月球引力对地震也有影响，有关专家已经并一直在研究。

4)地球人的精神生活

月亮对地球人的精神生活自古以来就扮演着很重要的角色，仅诗仙李白一人就有多篇吟月诗句。

《月下独酌》

花间一壶酒，
独酌无相亲。

举杯邀明月，
对影成三人。

《静夜思》

床前明月光，
疑是地上霜。
举头望明月，
低头思故乡。

披星戴月行路人，边疆、海岛战士们，月下歌舞儿女们、海洋作业军、民们……多少人在享受免费天灯赐予的温柔乳白的照明！环境设计师、建筑师、规划师、工程师们定能利用这盏免费天灯创造出更加美好的夜景。

things would be hard to live.

2) Tide effect

Moon's gravity makes earth sea water can't synchronously move to the direction of earth revolving on its own axis, the sea water and sea bottom move in opposite direction each other, just this action causes tides.

Using tide effect to store sea water to generate electricity this is a renewable sustainable energy.

3) Earthquake effect

Moon's gravity also influences earthquake. Relative specialists have been and are still studying the influence.

4) Earth man's cultural life

Since ancient time, for earth man's cultural life the moon has played an important role. Such as the Poetic immortal Li Bai alone had a lot of poems to sing the moon, examples:

《Drink Alone under the Moon》(引自英汉对照《唐诗三百首新译》)

Amid the flower, from a pot of wine

I drink alone beneath the bright moonshine.

I raise my cup to invite the Moon who blends her light with my shadow and we're three friends.

《Thoughts in the Silent Night》(引自英汉对照《唐诗选》)

Beside my bed a pool of light—
is it hoarfrost on the ground?
I lift my eyes and see the moon,
I bend my head and think of home.

Walkers under the canopy of the moon and stars, soldiers guarding at borderlands and islands, men and women singing and dancing under the moon, army and people of doing sea works…so much people have been enjoying

5) 有利于地球人户外夜生产与生活。

the smooth and milky lighting vouchsafed by the free sky lamp!

Environmental designers, architects, planners and engineers certainly can use the free sky lamp to create more better night environments.

5) Benefiting for earth man's outdoor night production and living.

※　　　※　　　※　　　※

问君有志太空否

地月浩瀚遥相邻

科技天神飞桥牵

问君有志太空否？
金帖相邀首站 moon,
娥、刚捧出桂花酒,
笑询扩建广寒宫,
备迎游客遨太空。

Are You Interesting to Go Space

A vast expance exists between the two neighbors earth & moon.

The Space God, science-technic can fly a bridge to link them.

Are you interesting to go space?

Gilt Card invites you first going to moon.

Chang'e and Wugang contribute the wine fermented with osmanthus flowers, and with smiling ask to expand the Moon Palace, so as to welcome tourists travelling through space.

1.2.2 地球

(1) 一般介绍

地球是地球人类起源、生存、发展的母体。图 1-1 是地球人血液中元素与地球岩石中元素关系对照表。

从图 1-1 可看出,地球人血液中的元素与地球岩石中元素的类型、含量竟达到如此惊人的同步波形!

地球表面约 67% 是水,人体重的 65%~70% 也是水,这组对应百分数,再次证明地球人与地球的密切关系。

1.2.2 Earth

(1) General

Earth is the matrix of earth man's origin, subsistence and development. Fig. 1-1 shows a comparison between the elements in man's blood and the earth rock. From Fig. 1-1 we can see the contents and types of the elements in human's blood and earth rock are so amazedly in step!

On earth, water has occupied about 67% of the total area of the ground, and there 65%~70% of the weight of a man is water. The two correspondent percents prove that again, between earth man and earth exists close relation. So, we

所以说:"地球是地球人最伟大的母体。"此话毫不为过。

say: "Earth is earth man's the greatest mother" that's no any exaggeration.

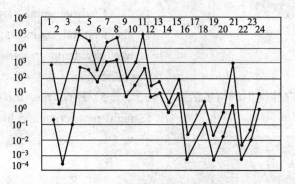

图 1-1 地球人血液元素与地球岩石元素对照表
Fig.1-1 Element quantity relation between in human blood and earth rock

1. 氢	2. 铍	3. 氟
4. 钠	5. 镁	6. 磷
7. 钾	8. 钙	9. 铬
10. 锰	11. 铁	12. 铜
13. 锌	14. 砷	15. 镓
16. 铑	17. 镉	18. 锡
19. 碲	20. 碘	21. 钡
22. 铼	23. 汞	24. 铝

1. hydrogen	2. beryllium	3. fluorine
4. sodium	5. magnesium	6. phosphorus
7. potassium	8. calcium	9. chromium
10. manganese	11. iron	12. copper
13. zinc	14. arsenic	15. gallium
16. rhodium	17. cadmium	18. tin
19. tellurium	20. iodine	21. barium
22. rhenium	23. mercury	24. aluminium

The element content of earth rock: $\mu g/g$;

The element content of human blood: $\mu g/ml$

(source: China Encyclopaedia 《Sanitary Science》)

天文学家研究,地球是太阳系中九大行星之一,太阳是银河系恒星之一。银河系有24500亿个恒星。宇宙在约150亿年前一次发生大爆炸,散向各空间的炽热质团在漫长的时间里逐渐散热冷缩,演变成千千万万亿如银河系的宇宙星系。

Under astronomer researches, earth is one of the nine big planets in solar system. The sun is one of the fixed stars in the Milky Way System (the Galaxy). There are 2450 billion fixed stars in the Galaxy. About 15 billion years ago, an universal Big Bang happened, the red-hot masses jetted to every direction. During the long years, the jetted masses gradually lost their heat and contracted and evolved as billions upon billions galaxies like the Milky Way System.

大约在66亿年前,银河系又发生一次大爆炸,其后16亿年有若干团星际物质散落在今天的太阳系空间,逆时针方向自转并公转

About 6.6 billion years ago, in the Galaxy happened again a Big Bang. Some 1.6 billion years later, several interstellar (intersteller)

运行,逐渐冷缩凝聚成太阳系的行星,地球即其中之一。太阳系行星系统较定形的年代约在46亿年前,故迄今天文界公认的地球年龄为46亿年。

地球赤道半径长为6378.245km,两极半径为6356.863km,平均半径为6371.2km,推算出地球总面积约为51000万 km^2,其中约71%为海洋面积。

中国陆地面积960万 km^2,约占地球陆地表面积14790万 km^2 的6.5%。中国陆地面积中,平原只有12%,各种山、坡、沟、洼地占70%以上。

(2)地球上其他物质环境
1)大气圈

地球表面上空有厚厚一层大气(空气)包围着,南、北两极区上空该厚度大于2800km,赤道上空该厚度大于4200km。

大气是混合体,其组成按干空气(不含水蒸气)体积百分比,分别为:氮78.08,氧20.95,氩0.93,二氧化碳0.03,以上四气体占总体积99.99%,其他微量气体为:氖1.8×10^{-3},氦5×10^{-4},氪1×10^{-4},氙1×10^{-5},臭氧约为1×10^{-6},氡6×10^{-18},氢$<1\times10^{-3}$。

含有上述空气组成成分及其体积百分比的空气即为新鲜空气(洁净空气)。

masses spread in the space of today's solar system, against the hour hand direction revolve round the sun and on their own axes, gradually cooled and contracted and evolved as the solar system's planets, earth is one of them. About 4.6 billion years ago, the shape of the planet system is rather stable, therefore, astronomic field universally acknowledge the earth age is 4.6 billion years till now.

The radius of earth equator is 6378.245km, of the two poles is 6356.863km, the mean radius is 6371.2km, and the computative total area of earth is about 510 million km^2, 71% of the total area water occupies.

Chinese land area is 9.6 million km^2, about 6.5% of the earth total land area of 147.9 million km^2. Within Chinese total land area, plain only some 12%, assorted (various) mountains, sloping fields, ravines and depressions occupy more than 70% of the total land area.

(2) Other physical environments on earth
1) Atmosphere

There is a very thick atmosphere around the earth ground. In the sky of Antarctic Region and Arctic Region the thickness is more than 2800 km, and of equator, more than 4200km.

Air is a gas mixture. According to the percentage by volume of dry air, the gas components includes respectively: (N)nitrogen 78.08, (O) oxygen 20.95, (Ar; A) argon 0.93, (CO_2) carbon dioxide 0.03, the above four gases occupy the air total volume 99.99%. Other microscale gases of the percentage by volume are: (Ne) neon 1.8×10^{-3}, (He) helium 5×10^{-4}, (Kr)krypton 1×10^{-4}, (Xe) xenon 1×10^{-5}, (O_3) ozone 1×10^{-6}, (Rn; Nt) radon (niton) 6×10^{-18}, (H) hydrogen

围绕地球的大气圈由对流层、平流层、电离层、散逸层等组成。

①对流层

厚：两极天空区 7~9km，赤道上空 15~17km。

质量：大气总质量约为 6 千亿吨，相当于地球质量的百万分之一。对流层质量约占大气总质量的 95%。

对流层的气流在垂直方向与水平方向对流运动均很明显。一年四季的气象变化，如：风、云、雷、电、雹、雨、雪、雾、露、霜均发生在此层内，有的如雾、露、霜则直接发生在接触地面或接近地面的空气里。

对流层发生的气象现象与地球生物关系最密切，其中有的如飓风与台风、厄尔尼诺与拉尼娜现象等都是严重的灾害。

②平流层（亦称同温层），含臭氧层

在对流层外到离地 50km 处大气中有一层只有水平对流的空气层，叫平流层。

在离地面 25km 以上的平流层内有一层厚约 30km 的臭氧（O_3）层 ozonosphere。

臭氧层是怎样形成的呢？

它是由太阳紫外线（ultraviolet rays）造成的：该紫外线将氧分子"O_2"分裂为氧原子"O"，随即有三个氧原子或一个氧分子与一个氧原子碰撞结合在一起，形成一个 O_3 臭氧分子，众多的 O_3 形成了约 30km 厚的臭氧层。正是太阳紫外线造成的这层臭氧层阻止

$< 1 \times 10^{-3}$.

The atmosphere around earth ground is composed of troposphere, stratosphere, ionosphere and dissipation sphere.

① Troposphere

Thickness: in the sky up the two Poles, 7~9km, in the sky up equator, 15~17km.

Mass: The total mass of the air is about 600 billion t, near one over one million of the earth total mass. The troposphere includes some 95% of the air total mass.

In troposphere, the horizontal and vertical convective movings are notable. The weather changing in four seasons of a year such as wind, cloud, thunder and lightning, hail, rain, snow, fog, dew and frost all occur in troposphere. Within them, fog, dew and frost occur on ground or in air near ground.

Weather phenomena occured in troposphere have the closest relation to earth living things, some of them such as hurricane and typhoon, Elnino and LaNina are heavily damaging disasters.

② Stratosphere (strato) including ozonosphere

In the space from outside of troposphere up to 50km from ground, only occurs horizontal convective air moving this air space is termed stratosphere.

In stratosphere up the 25km from ground there is an ozonosphere of about 30km thickness.

What causes the ozonosphere?

Solar ultraviolet rays cause the ozonosphere. The ultraviolet rays split oxygen molecule "O_2" into atoms "O", then three oxygen atoms or one oxygen molecule and one oxygen atom combine to form an ozone molecule "O_3" in collision. A great many ozone moleculae have formed the ozonosphere of

了太阳投向地球99%的紫外线,使地球生物免遭毁灭性紫外线杀伤危害。透过大气到达地面1%的太阳紫外线则起到消毒、灭菌、助长儿童骨骼生长(消除佝偻病)的有益作用。

保护臭氧层就是保护我们自己。

③电离层

地面以上50km～100km空间有一电离层,由太阳紫外线电离空气形成。电离层可反射无线电波传递无线电信息。

④散逸层或称外大气层

散逸层是最外层的大气空间,范围距地500km～65000km。空气极稀薄,其密度小于近地空气密度的300万亿分之一。

地球大气层还一直在保护地球免遭无数星体的撞击,使该星体在摩擦生热中成为碎陨石。

2)天灾人祸负效环境

天灾人祸造成的负效环境是围绕人类的环境重要组成部分,对环境破坏极大。

a)人为负效环境

①破坏绿色世界,加剧大气温室效应

通过人类生产、生活及动植物呼吸作用大量排放到大气中的二氧化碳、水蒸气等有类似普通玻璃那样"透短吸长"的热学特性,

about 30 km thickness. Just this ozonosphere caused by the solar ultraviolet rays has stopped 99% of the solar ultraviolet rays projecting on earth, and protected earth from exterminatory kill of the ultraviolet rays. The remaining 1% ultraviolet rays through atmosphere reached ground are good rays, benefit sterilizing, killing germs and helping children's bones growing (avoiding rickets).

Protecting ozonosphere is protecting ourselves.

③ Ionosphere

Above ground 50km～100km, there is an ionosphere caused by solar ultraviolet rays. Ionosphere can reflect radio waves and pass radio information.

④ Dissipation sphere or outer atmosphere

This sphere is the outmost space of atmosphere, its confines are above ground 500km to 65000 km, here the air is very rarefied, its density is less than one over 3×10^{12} of that of the air near ground.

Atmosphere has still protected earth from bumpings of no end of the celestial bodies, made them heated by friction and become into meteoric blocks or chippings.

2) Negative effect environments by man and natural calamities

Negative environments by man-made and natural calamities are also an important component of the environments around human being, which have damaged the environments very seriously.

a) Negative environments by man

① Damaging green world, aggravating atmospheric greenhouse effect

There have been much carbon dioxide and water vapour, etc. exhausted to atmosphere by man's producing, living and breathing out

可透过太阳向地面的辐射(属短波辐射),同时又吸收地面向空间的辐射(长波辐射),使大气升温,并以长波辐射方式将大部分热量返回地面,使地球升温,这就是通称的大气温室效应。排放到大气中的二氧化碳增多是由于人们不断破坏绿色世界的结果。地表森林及其他绿色植物以及两极冰山是稳定地球气候及生态平衡的两大力量,科学家们分别称为"绿色力量"和"白色力量"。但由于滥伐森林,侵占绿色面积(建筑业是最大的侵犯户),使植物进行光合作用、释放氧气、吸收二氧化碳的性能大减。相反人类生产、生活及动植物呼气排放到大气的二氧化碳又不断增加致使大气温室效应加剧,引起地表热胀变形,山、川、路、房受害,海水热胀升高,加之两极冰山消融进一步升高海洋水位,最高可升高60m以上,严重危及沿海城乡生存。

滥伐森林,破坏绿色世界还会导致严重水土流失,加剧洪水灾害与生态失衡。

from animals and plants. These carbon dioxide and water vapour have a thermal character of "transmitting short wave radiation while absorbing long wave radiation"like glass does, they can transmit solar radiation (short wave radiation) reaching ground while absorb the ground radiation (long wave radiation) and cause atmosphere temperature higher, emitting most heat back to ground by long wave radiation form, and causing ground temperature higher, that's generally termed "atmosphere greenhouse effect."

Carbon dioxide has been exhausted to atmosphere more and more because the green world has been damaged by man more and more. Ground forests and other green plants and the two Poles icebergs are two great forces for stabilizing earth climate and eco-balance, scientists term the two forces as "Green force" and "White force" respectively. However, much forests denuded, much green area seized (building trade is the biggest aggressor) cause plant photosynthesis and the ability of releasing O_2 and absorbing CO_2 seriously decreased. On the other hand, the CO_2 exhausted to atmosphere more and more by man's producing, living and breathing out from animals and plants, which certainly aggravates atmosphere greenhouse effect, and causes that: ground expanding to damage mountains, rivers, roads and houses; sea level higher and icebergs melted to further rise sea level, final the sea level might be risen more than 60m which would seriously damage the cities, towns, villages in the boastlands.

Forests denuded and green area seized can also cause badly soil erosion, aggravate flood disaster and eco-balance loss

②破坏臭氧层

前已述及,在大气平流层内有一层臭氧层是保护地球及地球生物免遭太阳紫外线伤害的保护伞。但长期以来人们使用的空调、冰箱、汽车、计算机以及发泡剂、清洗剂等含有氯氟烃类物质,排放到大气中,在太阳紫外线照射下,氯氟烃分子会裂解,放出游离的氯原子,氯原子很活跃,它会从臭氧分子"O_3"中拉出氧原子"O",臭氧分子就减少(一个氯原子大约可破坏 10 万个臭氧分子)。臭氧浓度就会不断降低,以致形成"空洞"(臭氧比正常少了 75%)。地球南、北极均已发现臭氧空洞,南极臭氧空洞已大于中、美两国陆地之和,并以每年一个中国或美国陆地面积那么大扩展。虽然许多企业已有改进措施,但长期积存在大气中的危害气体还会长期地起破坏作用。

③人为水污染

在中国,生产、生活排放的污水已使全国 80% 的地表水与地下水受到污染。

④人为大气污染

我国每年生产、生活、运输排放大量有害物质如二氧化碳、烟尘、工业粉尘、加上全国 2 亿 5 千多万户的炊烟致使我国大气受到严重污染。全国城市中仅 1/3 人口可吸到新鲜空气。

② Damaging ozonosphere

As has said before, in the stratosphere, there contains an ozonosphere, a protective umbrella to protect earth and its living things against the damage of solar ultraviolet rays. But the long used air conditioners, ice boxes, autocars, computers and vesicatory, abstergent, etc. contain hydrochlorinefluor which has been exhausted much to atmosphere, under solar ultraviolet ray's action, this gas is splited and release dissociated chlorins. These active chlorins can draw oxygen atom(O) from ozone molecule (O_3) and reduce ozone moleculae, (one chlorin atom can break about 0.1 million ozone moleculae) while make ozone consistency lesser and lesser and form "ozone hole" (ozone consistency is 75% lesser than the normal consistency).

In the South Pole and North Pole, ozone holes have been found. In South Pole the ozone hole's area has been bigger than the land of China and America together, and every year is still spreading with a area like one Chinese land or one American land. Although many trades have been doing improvements, the harmful gas long having been stored in the air will do damage for long.

③ Water polluted by man

In China, 80% surface water and underground water have been polluted by the waste water drained off from producing and living.

④ Air polluted by man

In China every year much harmful matters have been exhausted into air, such as carbon dioxide, smoke, dust by producing, living and traffic, in addition, there are more than 0.25 billion homes' cooking smoke exhausted into air. The above two together have made the air seriously polluted. Only 1/3

⑤人为火灾

据记录，我国每年人为火灾不下10万起，例如2000年全国就发生18万多起人为火灾，损失258亿元人民币。

⑥核危害

人们绝难忘记1945年日本广岛、长崎两城市被两颗原子弹夷为平地，30万人惨遭杀害！更多的人受到核辐射伤害！人们也绝难忘记前苏联切尔诺贝利核电站1984年4月的核泄漏悲剧，14万人受到严重核辐射伤害，大范围生态受到严重破坏。

长期的军备竞赛，我们的地球已经遭受了几百次核爆炸污染！21世纪核危害加剧的隐患一直存在。

⑦战争

自古以来，特别是冷兵器变为热兵器以来，战争是侵害人民生命财产、毁坏建筑及其环境、破坏生态与持续发展的最大祸首。

⑧恐怖分子的危害

⑨1966~1977"文化大革命"对我国各方面均造成了严重的危害。

⑩其他人为危害

吸毒在全球成亿计地夺去人的躯体、意志、家庭。有些吸毒者还会进行盗、骗、抢、杀等犯罪行为。

people through China can breathe in fresh air.

⑤ Fires by man

According to records, in China every year more than 0.1 million man-made fire cases occured, such as in 2000, there were more than 0.18 million fires by man, 25.8 billion Yuan RMB lost.

⑥ Nuclear damage

People never forget that in 1945, the two cities: Fukuyama and Nagasaki of Japan were levelled to the ground by two nuclear bombs, 0.3 million people miserably killed! much more people injured (harmed) by nuclear radiation. People also never forget that in former CCCP there occured a tragedy of nuclear leak at Ternopol nuclear power station, 0.14 million people were seriously injured by nuclear radiation, and the eco-environment in a large area badly damaged.

During the long time armament rivalry (rivalship), our earth has been suffered the pollutions of nuclear blasts for hundreds times! In the 21st century the nuclear damage might be aggravated, this covered bane exists all along.

⑦ Wars

Since ancient times, especially since cold weapon became hot weapon, war has been the biggest chief culprit in killing people, encroaching properties, destroying buildings and environments and damaging eco-sustainable development.

⑧ Terrorists' damages

⑨ 1966 ~ 1977 "The Great Cultural Revolution" has seriously damaged all fields in China.

⑩Other harms by man

Drug taking has deprived hundreds of millions of men's bodies, volition and families through the world. Some druggers might do

艾滋病在全球正以每天7000人的速率传染。我国已发现8种类型艾滋病感染者，人数已超1000万，多为青少年。

其他性病我国已有600多万人感染。

贪污腐化自古以来就是对为官担职人群中意志薄弱者的恶性污染。我国建筑业中贪污腐化是最严重者之一。

上述人为负效环境已使社会环境、经济环境、政治环境、精神环境均已受到严重污染。

b)天灾负效环境

①洪水

全球每年洪水都要使几亿人受害，几千万间房屋受损，几百亿美元经济损失。

中国1998年特大洪水为主的自然灾害中，3.5亿人次受害，死4000多人，塌房558万间，损房1205万间。经济损失3000多亿元。

②地震

地震可以说天天有。20世纪地震死亡人数已逾百万，一次死10万人以上的有4次（中国占2次）。最惨痛的是1976年7月28日中国唐山8级大地震，死24万2千人，伤14万多人，全城房屋除极少数外几乎全都倒塌，财产损失100多亿元。

crimes such as pilfering, diddling, snatching and killing.

AIDS has been infectioning sufferers with the rate of 7000 persons per day. In China have been found sufferers of eight types of AIDS. The sufferers have been over ten million persons, most of them are teenagers(youngsters).

In China, the sufferers of other venereal diseases (V. D) have been exceeding six million personae.

Corruption since ancient times has been still a vicious pollution to the weak-willed officers. Corruption in building trade is one of the most serious ones through China.

The above man-made negative environments have made social environment, economic environment, political environment and psychic environment polluted seriously.

b) Negative environments by natural disasters

① Floods

Floods through the world every year jeopardize hundreds of millions people, destroy tens of millions houses and make tens of billions USD lost in economy.

In 1998, China had a king-size(outsize) flood, 0.35 billion person-time damaged, more than 4000 people killed, more than 5.58 million houses collapsed, more than 12.05 million houses damaged, more than 300 billion Yuan RMB lost in economy.

② Earthquakes

Earthquakes occur every day. In 20th century there were more than one million people killed by earthquakes. Four of those earthquakes were more serious, in each of the four, more than 0.1 million people killed(two of the four occured in China), the deepest grieved one occured on July 28th 1976, in

③ 厄尔尼诺与拉尼娜

厄尔尼诺是一种影响全球的灾难性气象现象。每隔 2～7 年,12 月左右(圣诞节前后),南太平洋(正处于夏季)不按通常吹东南信风,而是反常吹南偏西风,把暖海水向东推,使赤道北的东太平洋含大量食料的深层冷海水不能上翻,导致大量浮游生物死亡,并引起连锁反应,使大量鱼群死亡、鸟类死亡或迁徙,肥料减少,农业减产……与此同时,大气从该温暖的海洋区得到高温水蒸气就更多,使气温升高,造成该大气所经地区不降雨率增多,形成旱灾。这股高温、高湿气流由于大气环流及地球自转的影响,流到较冷区,又造成过量降雨,形成洪涝灾害,这就叫厄尔尼诺现象。其间,在澳大利亚的悉尼与阿根廷的布宜诺斯艾利斯(同处南纬 35°附近)之间大气气压产生此高彼低、此低彼高的儿童玩跷跷板现象称南方涛动现象,对厄尔尼诺会起加剧作用,乃合称 ENSO 现象或 ENSO 事件。

Tangshan City, Northern China, magnitude 8.0, 0.242 million people killed, more than 0.14 million people injured, almost all houses in the city collapsed, and more than 10 billion Yuan RMB lost in economy.

③ EL Nino and La Nina

EL Nino is a weather disaster phenomenon affecting whole world. Every 3 to 8 years, about in December, round about the Christmas Day, the South Pacific (just in summer) does not blow normal trade SE wind, instead blows anomalous south by west wind, this wind pushes the South Pacific warm sea water to the East Pacific on the north of equator, the warm sea water causes there bottom cold sea water containing abundant foodstuff can't upturn to the sea surface, results in much planktons died, and in chain reaction, much shoals died, much aves died or moved off, fertilizer reduced and agriculture dropped in production... While atmosphere gains more high temperature water vapour from the heated sea water, causing rise of air temperature. The air with high temperature and high humidity reduces raining chance in the region where the air flow blows over, and causes drought there. Under the affection of atmospheric circulation and earth revolving on its own axis, the above air flow of high temperature and high humidity moving to cold region causes heavy raining forming flood/waterlogging disasters. This phenomenon is called EL Nino Phenomenon.

During EL Nino phenomenon, between Sydney, Australia and Buenos Aires, Argentina (the two cities are located at near south latitude 35°S), the atmosphere pressure has a phenomenon of that as one falls, another rises as children play on a seesaw, this is called

拉尼娜现象则是由于表层海水温度反常地降低引起的与厄尔尼诺现象方向相反的旱涝灾害,故又称反厄尔尼诺现象,1998年,由于厄尔尼诺与拉尼娜灾难气候,已造成40多国旱灾,20多国水灾,约三亿人无家可归,890亿美元经济损失。

④台风,飓风(龙卷风)

孟加拉一次强台风,风速72m/s,13万多人死去;美国是多飓风国家,飓风一般在12级以上,≥32.7m/s。美国佛罗里达州一次飓风将260万km^2地区房屋夷为平地,经济损失300多亿美元。

我国每年都会受到多次台风袭击,造成沿海地区巨大灾害。

⑤其他自然灾害

雷击(温度可达25000℃)、森林火灾、冰雹、火山喷发、滑坡、沙漠化等造成的负效环境也是发展可持续环境的巨大障碍。

1.3 房屋及其环境是如何建起来的?

两大工序:设计→施工,以房屋为例:

1.3.1 设计分类

(1)建筑设计,其主要任务是:
1)划分建筑空间,组织房间类型、数目与

South Oacillation, it aggravates EL Nino, their combination called ENSO.

La Nina phenomenon is opposite to EL Nino, due to the sea surface water temperature anomalously dropping down which causes drought/waterlogging disasters on the direction opposite to EL Nino, so, La Nina is called anti-EL Nino phenomenon.

In 1998, EL Nino and La Nina caused droughts in more than 40 countries, floods in more than 20 countries, 0.3 billion people homeless, 89 billion USD lost in economy.

④ Typhoon, hurricane

In Bangladesh, there got a typhoon, wind speed 72m/s, more than 0.13 million people killed; in America frequently occur hurricanes, often > 12th grade, ≥32.7m/s. In Florida, a hurricane razed all the houses. in a region of 2.6 million km^2, more than 30 billion USD lost in economy.

China, every year suffers more-typhoon raids causing coastal areas serious disasters.

⑤ Other natural disasters

Strike by lightning (temperature may reach 25000℃), forest fire, hailstone, volcanic eruption, landslide, desert spread, etc. have caused negative effect environments, they are also big obstacles for developing sustainable environments.

1.3 How to Build a House and Its Environments?

Two big processes: Design → Construction, take a house as an example:

1.3.1 Design classification

(1) Building design, main jobs:
1) Dividing building spaces, organising

层数,确定体形;

2)建筑技术设计:

①传统概念:从使用功能与围护的角度出发,确定相关构件的功能、构造原理、方法、材料。例如屋顶与外墙,其主要功能是分隔室内外空间并围护房屋及居民免遭水(降雨、排水、地下水)、火、风、雪、冰、雹、沙尘暴、噪声、污染、冬冷夏热以及地震等侵害。

②生态可持续建筑概念:建筑构件应成为可持续构件,例如"散水",传统观念的散水只是将天空、屋面和墙面降到散水上的雨水排走以保护地基与基础免受水的侵害。但散水雨天将雨水排走晴天又反射浪费大量太阳能,从全国、全世界来看,确是对物质流与能流的极大浪费。因此传统的建筑散水不是可持续构件。

将散水进行绿化种植,则不仅可利用雨水与太阳能,而且更有效地保护了地基与基础,同时还可获得良性的生态效应与美化的环境。这种构件在21世纪乃至将来都值得发展,是可持续构件。

3)建筑装饰(表面处理)与环境设计

(2)结构设计,其主要任务是:

1)选择承重构件的类型与材料;

(organizing) the room types, scalar and storeys (stories) of the building and fixing on the building shape;

2) Architectural technic design:

① Conventional concept: Based on the use and protection, designers should fix on the functions, constructional principle, means and materials of elements concerned. Examples: Roof and outer walls, their main jobs are to isolate indoor space from outdoor space and to protect the house and dwellers from damages of water (rainwater, drainage and underground water), fire, wind, snow, hailstone, dust storm, noise, pollution, winter cold, summer hotting and earthquake, etc.

② Eco-sustainable architecture concept: Building elements should become sustainable elements, such as "apron", in conventional concept, apron's function is only to drain the rainwater from sky, roofing and walls so as to protect ground base and foundation from damage of the water. But the apron drains the water during raining days, also wastes much solar energy by reflection during fine days. Through whole country and world, so much mass flow and energy flow have been wasted by aprons that conventional building aprons are not sustainable elements.

Planting flowers on apron that not only can utilize the rainwater and solar energy but also better protect ground base and foundation, at the same time we can get a good ecologic effect and beautiful environment. This element is valuable to be developed in 21st century and further future, it's a sustainable element.

3) Building decoration(surface treatment)and environment design

(2) Structural design, main jobs:

1) Selecting types and materials of load-

2)分析、计算承重构件并进行配料。

(3)其他专业设计,如给排水,供热、制冷、通风、空调、照明等。

(4)城市规划设计,是人聚环境最主要的设计之一。

1.3.2 施工的主要任务

(1)按设计图组织施工,首先与上述设计者开好技术交流合作会;

(2)做好施工组织设计;

(3)做好施工技术与施工进度设计。

1.3.3 简言之

建筑物及人工环境以及建筑群及其自然环境组成的乡、镇、城市乃是脑力与体力劳动集体协作的产物。

建房或建造环境可以没有图纸,但必须有设计。即使造一间鸡舍,你必须想到场地在何处、什么材料及外形如何等,这种考虑就是设计。

人类建房与动物造舍有何不同呢?

其根本区别就是:人类建房依靠设计并用工具施工,而动物(蜂、蚁、鸟……)建舍靠的是本能(迄今人类所知)。

bearing elements;

2) Analyzing and calculating the bearing elements and batching for them.

(3) Other professional designs such as water supply and drainage, heating and cooling, ventilation and air conditioning and lighting, etc.

(4) Urban planning design is one of the most important designs of people living environment.

1.3.2 Construction main jobs

(1) Organizing construction according to design drawings, first holding well the meeting with the above designers for technologic exchang and cooperation;

(2) Doing well construction organization design;

(3) Doing well construction technology and progress designs.

1.3.3 In brief

Buildings and built(man-made) environments and they form the villages, towns and cities are the products completed by collective cooperation of many different brainworks and physical (manual) labours.

To build a house or an environment, you may have no any drawing, but you must have a design. For an example, even build a hens' nest you must think where's the site, what are the materials and how's the shape? This 'thinking' is right the design.

What is the difference between man-built houses and animal-built nests?

The root difference between man‐built houses and animal‐built nests is that man builds houses according to design and with tools, but animals (bees, ants and birds...). build nests only by their instinct (by man's knowledge till now).

1.3.4 如何感知房屋及其环境？

房屋及其环境所以被感知乃由于人体与房屋及其环境之间的信息交流。人体的感官有眼、耳、鼻、舌、身、皮、脑。

就理性分析而言,我们从未真正见到过"材料",我们所感知的只是光的反射或折射,所以我们必须懂得光学,选好材料并做好表面处理。同理我们必须懂得声学、热学、建筑构造学、生态学以及可持续发展原理等,才能干好本专业工作。

建筑师、规划师、环境设计师以及艺术家必须博学多才。

1.4 建筑物分类,分等及防火分级

1.4.1 建筑物分类

(1)依据主要承重材料分：
1)土结构
2)木结构
3)石结构
4)混合结构：
①砖木混合结构
②砖混结构,即砖—钢筋混凝土结构
③钢混结构即钢与钢筋混凝土结构
④竹木结构
⑤石木结构
5)钢筋混凝土结构：

①钢筋混凝土框架结构

1.3.4 How to experience houses and their environments?

Houses and their environments are experienced by an exchange of information between the body and houses and environments. The sensors of body are eyes, ears, nose, tongue, body, skin and brain.

In an analytical sense, we never really "see material", what we perceive are reflections and refractions of light. So we must understand light and well select materials, well treat the surfaces. Also we must understand sound, heat, building techtonics, ecology and sustainable developing principles &c. We can then well do our professional jobs.

Architects, planners, environmental designers and artists must be learned and versatile.

1.4 Types, Ratings and Fireproof Classes of Buildings

1.4.1 Types of buildings

(1) According to main bearing materials:
1) Soil (earth) construction
2) Timber (wood) construction
3) Stone construction
4) Composite construction:
① Wood-brick construction
② R.C. - brick construction (R.C. = reinforced concrete)
③ R.C.-steel construction
④ Timber-bamboo construction
⑤ Timber-stone construction
5) Reinforced concrete construction (R.C. construction)
① R.C. frame construction

②钢筋混凝土薄壳结构
③钢筋混凝土拱结构等。
6)钢结构等
(2)按用途分：
1)民用建筑,如：住宅、商店、学校、办公楼、博物馆、展览馆、图书馆、旅馆、医院、体育馆、剧院、电影院等；

2)工业建筑,如各种生产车间等；
3)农业建筑,如农业温室等；

4)军用建筑等。
(3)按施工分：
1)现场建造,如现浇钢筋混凝土建筑；

2)预制装配,如装配式钢筋混凝土房屋；

3)现场建造与预制装配相结合。

(4)按层数分：
少层建筑；多层建筑；高层建筑。

我们为什么要对建筑物进行分类呢？

(1)取个名字,建立概念,便于用术语交流；
(2)便于分门别类,深入研究发展建筑科学。

什么叫"科学"？科学就是分门别类去研究。如医学中的内科、外科、妇科、儿科……建筑及其环境领域的建筑设计、城市规划、环境艺术等,"学"就是研究,分门别类就有利于更深的研究,求得更高的发展。

② R.C. shell construction
③ R.C. arch construction, etc.
6) Steel construction, etc.
(2) According to uses:
1) Civil architecture such as: dwellings, shops, schools, office buildings, museums, exhibition buildings, libraries, hotels, hospitals, gyms(gymnasiums), theatres and cinemas, etc. ;
2) Industrial buildings such as workshops, etc.
3) Agricultural buildings such as agricultural greenhouses &c.
4) Military buildings &c.
(3) According to construction:
1) Constructing in site such as a cast-in-place R.C. building;
2) Prefabricated construction such as a precast R.C. house;
3) Combined with Constructing-in-site and prefabricated parts
(4) According to stories:
Less-story buildings, multi-story buildings and high-rise buildings.
Why do we divide buildings into different types?
（1）Naming a name, setting concepts, using technics(terms) so as to easy exchange;
（2）Benefiting individually to deep study and develop architectural science.
What means "science"? "Science" means dividing an object of study into relative branches/professions to study such as in medical medicine, internal medicine, surgery, gynecology, pediatrics (paediatrics)... in architectural and its environmental fields: building design, urban planning and environmental art, etc. "study" is researching, dividing an object of study into relative branches benefits to deeper study them and higher develop them.

最后两点：

(1)建筑分类的名称是可变的，如："木结构"只突出材料；"现浇钢筋混凝土框架办公楼"则同时说明了施工、材料、结构与用途。

(2)建筑分类项目是发展的，随生产、科技、生活水平的提高而发展。如：无毒塑料结构、太阳能建筑、生态建筑，可持续建筑，纳米材料建筑等。

Final two points:

(1) The names of building types are alterable (variable), such as "timber construction", it only explains that the building main material is wood; "a cast-in-site R.C. frame office building" that not only explains the building main material is reinforced concrete but also the building structure is R.C. frame, the construction method is cast-in-site and the use of the building is for office;

(2) The items of building types are developing following the progresses of producing, science - technology and living level, such as poisonless plastic structures, solar buildings, ecologic buildings, green architecture, sustainable architecture and Nami material buildings, etc..

1.4.2 建筑物分等(表1-1)

1.4.2 Ratings of buildings (Table 1-1)

表1-1 按耐久性规定的建筑物等级

建筑等级	建筑物性质	耐久年限
一	具有历史性、纪念性、代表性的重要建筑物，如纪念馆、博物馆、国家会堂等	100年以上
二	重要的公共建筑，如一级行政机关办公楼、大城市火车站、国际宾馆、大体育馆、大剧院等	50年以上
三	比较重要的公共建筑和居住建筑，如医院、高等院校，以及主要工业厂房等	40~50年
四	普通的建筑物，如文教、交通、居住建筑以及工业厂房等	15~40年
五	简易建筑和使用年限在十五年以下的临时建筑	15年以下

· 耐久性
· 耐久年限

· durability
· duration of the building

Table 1-1 Building Ratings Based on the Durability

建筑等级 Building Ratings	建筑物性质 Building Properties	耐久年限 Durable Age(years)
一 1st	The important buildings with historicity, memorability or typicality such as memorial halls, museums and national hall, etc.	>100
二 2nd	Important public buildings such as the first class administration offices, big city train stations, international hotels, big gyms and big theaters, etc.	>50
三 3rd	Rather important public buildings and dwelling houses such as hospitals, colleges and universities and main industrial factories	40~50
四 4th	Common buildings such as culture & education buildings, traffic buildings and industrial factories, etc.	15~40
五 5th	Buildings simply constructed and use age less than fifteen years	<15

1.4.3 建筑物防火分级

建筑物防火分级及构件燃烧性能和耐火极限见表1-2。

"建筑构件耐火极限(也称耐火时限)"术语解释：

建筑构件按国际标准化组织的标准进行试验,其耐火时限的含义是：

(1)构件单面受火作用开始到失去支承能力所经历的小时数；

(2)或到发生穿透裂缝所经历的小时数；

(3)或到背面温度达到220℃所经历的小时数。

1.4.3 Fireproof classes of buildings

Building fireproof classes, and member combustibility and Fire Resistance Hour (FRH) see Table 1-2.

An explain on the term of "Fire resistance hour(FRH) of building elements":

According to the standards of the International Standard Organization (ISO) to test the building elements, the meaning of Fire Resistance Hour(FRH) is that：

(1) The undergone hours from the beginning of firing one side of the element till its bearing capacity lost；

(2) or till penetrated crack appeared；

(3) or till the temperature of the element reverse up to 220℃.

表 1-2　建筑物耐火等级、建筑构件的燃烧性能和耐火极限(时限)

构件名称		燃烧性能和耐火极限(h) 耐火等级	高层建筑 一级	高层建筑 二级	普通建筑 一级	普通建筑 二级	普通建筑 三级	普通建筑 四级
墙	防火墙		非燃烧体 3.00	非燃烧体 3.00	非燃烧体 4.00	非燃烧体 4.00	非燃烧体 4.00	非燃烧体 4.00
	承重墙、楼梯间墙		非燃烧体 2.00	非燃烧体 2.00	非燃烧体 3.00	非燃烧体 2.50	非燃烧体 2.50	难燃烧体 0.50
	电梯井和住宅单元之间的墙		非燃烧体 2.00	非燃烧体 2.00				
	非承重外墙、疏散走道两侧的隔墙		非燃烧体 1.00	非燃烧体 1.00	非燃烧体 1.00	非燃烧体 1.00	非燃烧体 0.50	难燃烧体 0.25
	房间隔墙		非燃烧体 0.75	非燃烧体 0.50	非燃烧体 0.75	非燃烧体 0.50	非燃烧体 0.50	难燃烧体 0.25
柱	支承多(高)层的柱		非燃烧体 2.50	非燃烧体 2.00	非燃烧体 3.00	非燃烧体 2.50	非燃烧体 2.50	难燃烧体 0.50
	支承单层的柱				非燃烧体 2.50	非燃烧体 2.00	非燃烧体 2.00	燃烧体
梁			非燃烧体 2.00	非燃烧体 1.50	非燃烧体 2.00	非燃烧体 1.50	非燃烧体 1.00	难燃烧体 0.50
楼板			非燃烧体 1.50	非燃烧体 1.00	非燃烧体 1.50	非燃烧体 1.00	非燃烧体 0.50	难燃烧体 0.25
屋顶承重构件			非燃烧体 1.50	非燃烧体 1.00	非燃烧体 1.50	非燃烧体 0.50	燃烧体	燃烧体
疏散楼梯			非燃烧体 1.50	非燃烧体 1.00	非燃烧体 1.50	非燃烧体 1.00	非燃烧体 1.00	燃烧体
吊顶(包括吊顶搁栅)			非燃烧体 0.25	难燃烧体 0.25	非燃烧体 0.25	难燃烧体 0.25	难燃烧体 0.15	燃烧体

Table 1-2 Building Fire Resistance Rating(FRR), Building Member Combustibility and Fire Resistance Hour(FRH)

燃烧性能和耐火极限 Combustibility &FRH (h) 构件 Members	耐火等级 FRR	高层建筑 High-rise Buildings		普通建筑 Common Buildings			
		一级 1st	二级 2nd	一级 1st	二级 2nd	三级 3rd	四级 4th
防火墙 Fire walls		非 Non 3.00	非 Non 3.00	非 Non 4.00	非 Non 4.00	非 Non 4.00	非 Non 4.00
承重墙、楼梯间墙 Bearing walls; Staircase walls		非 Non 2.00	非 Non 2.00	非 Non 3.00	非 Non 2.50	非 Non 2.50	难 Nonf 0.50
电梯井；住宅单元之间的墙 Lift well walls; Walls between dwelling units		非 Non 2.00	非 Non 2.00	/	/	/	/
非承重外墙；疏散走道两侧的隔墙 Nonbearing outer walls; Partitions at both sides of escape walkway		非 Non 1.00	非 Non 1.00	非 Non 1.00	非 Non 1.00	非 Non 0.50	难 Nonf 0.25
房间隔墙 Partitions between Rooms		非 Non 0.75	非 Non 0.50	非 Non 0.75	非 Non 0.50	非 Non 0.50	难 Nonf 0.25
支承多(高)层的柱 Columns of Multi-story or High-rise Buildings		非 Non 2.50	非 Non 2.00	非 Non 3.00	非 Non 2.50	非 Non 2.50	难 Nonf 0.50
梁 Beams		非 Non 2.00	非 Non 1.50	非 Non 2.00	非 Non 1.50	非 Non 1.00	难 Nonf 0.50
楼板 Floors		非 Non 1.50	非 Non 1.00	非 Non 1.50	非 Non 1.00	非 Non 0.50	难 Nonf 1.25
屋顶承重构件 Bearing Members of Roof		非 Non 1.50	非 Non 1.00	非 Non 1.50	非 Non 0.50	燃 Comb	燃 Comb /
疏散楼梯 Escape Stairs		非 Non 1.50	非 Non 1.00	非 Non 1.50	非 Non 1.00	非 Non 1.00	燃 Comb /
吊顶(含吊顶搁栅) Suspensory Ceilings(incl. Ceiling joists)		非 Non 0.25	难 Nonf 0.25	非 Non 0.50	难 Nonf 0.25	难 Nonf 0.15	燃 Comb /

Note: In the Table, 非____、Non ____ = 非燃烧体 Noncombustibles; 难____、Nonf = 难燃烧体 Nonflammables; 燃____、Comb____ = 燃烧体 Combustibles

1.4.4 建筑物组成构件

建筑物组成构件见图 1-2,
基础

1.4.4 Components of a building

Components of a building see Fig. 1-2.
foundation

1　环境·人·建筑　　23

阳台	balcony
散水	apron
楼板	floor
台阶	a flight of steps; steps leading up to a house; steps
圈梁	girt(girth)
地面	ground
内纵墙	inner lengthways wall
雨棚	canopy
内横墙	inner cross wall
楼梯段与平台板	stair flight/landing
檐沟	eaves gutter
门、窗	door/window
屋面板	roof slab

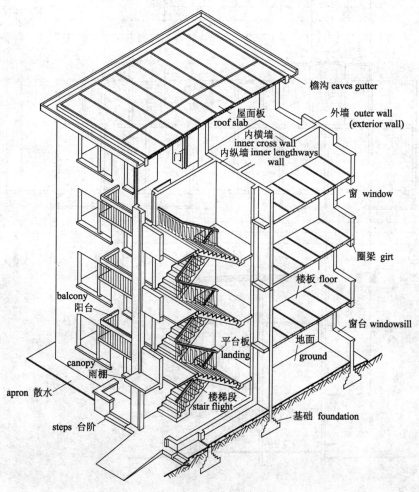

Fig.1-2　Components of a building
图 1-2　建筑物组成构件

1.4.5 建筑物平、立、剖面表示法(图 1-3)

1.4.5 Showing of a building plan, facade and section(Fig. 1-3)

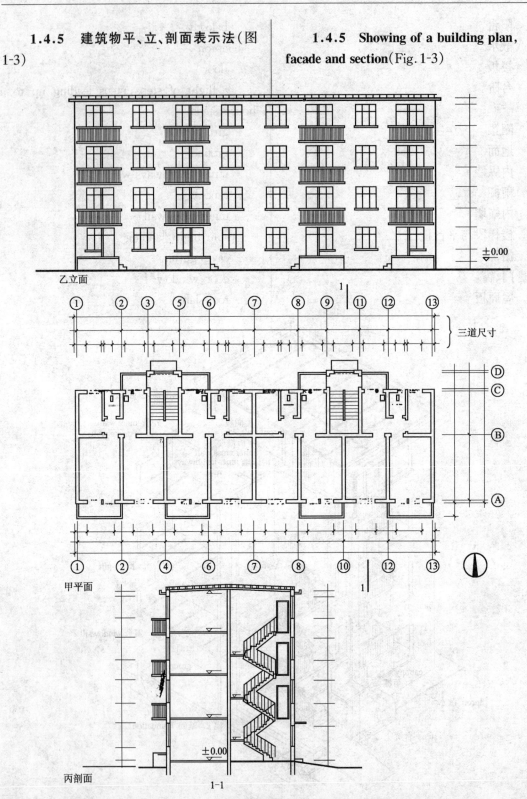

Fig. 1-3 A showing of a building plan, facade and section

图 1-3 建筑物平、立、剖面表示法

2 建筑科学基础

2 FOUNDATIONS OF ARCHITECTURAL SCIENCE

2.1 地基

定义：承受基础传来的荷载而产生应力应变的土层叫地基。

地基分天然地基与人工地基：

（1）凡天然土层具有足够的承载力，能安全承受房屋荷载的地基称天然地基。

（2）当土层承载力不够，需经人工加强承载力的地基称人工地基。

天然地基与人工地基的概念是相对的。同一地基，对于轻荷载房可以是天然地基，对于重荷载房则需处理成人工地基了。

人工地基的主要方法有夯实、压实、换土和打桩等（表2-1）。

图2-1 为地基受基础传来荷载后的压力分布状况。

2.1 Ground Bases

Definition: The soil layer having stress and strain caused by the load transmited by the foundation is called ground base.

Ground bases can be divided into subsoil (natural ground base) and man-made ground base.

(1) Any natural soil layer having enough bearing capacity to safely bear the building load is termed subsoil.

(2) A soil layer having not enough bearing capacity which needs man-made reinforcement is termed man-made ground base.

The concept of subsoil or man-made ground base is relative. A same ground base, for a light-weight building may be a subsoil, for a heavy-weight building, it may need to be treated as a man-made ground base.

The main methods of man-made ground bases are ram, compaction, soil changed and piling (Table 2-1).

Fig.2-1 Shows in ground base the pressure distribution under the loads transmited by the foundation.

表 2-1 常用人工地基举例
Table 2-1 Examples of General Man-made Ground Bases

方法 Methods	说明 Explanation
蛙式夯 Frog Rammer	重(Weight)50~60kg 落高(Fall-height)0.5~0.6m, 加固深度(Consolidating depth)0.2~0.25m
重锤 Heavy Hammer	锤重(Hammer weight)1~1.5t 锤直径(Hammer diameter)1m 落 高(Free-fall height)3~4.5m 加固深度(Consolidating depth)1.2m
压路机滚压 Rolling by Roller	每层铺土(Each course of earth paved)0.3~0.5m 压路机碾压(Rolled by roller)4~12遍
振动板夯实机 Vibrating Plate Compactor	自重(Dead weight) 2t 振频(Vibrating frequency)1100~1200转(circles)每分钟(Per minute)
换土 Earth Changed	必要时根据设计用好土换掉坏土。 If necessary, according to design to change the bad earth for good one.

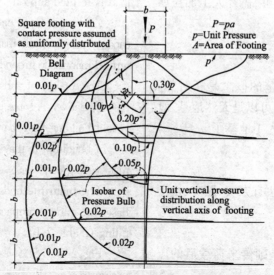

Fig.2-1　Relation between bell diagrams and bulb of pressure
图 2-1　钟形图与压力泡的关系

2.2 基础

定义:建筑物最下面埋入土中的部分叫基础。

2.2.1 基础的类型

(1)按地基承压面及基础几何形状(见图2-2)

1)单独基础;
2)条形基础或带形基础含双向带形基础;
3)筏形基础(浮筏式基础、满堂基础);
4)钢筋混凝土箱形基础。

(2)按基础的力学性能

1)刚性基础(图2-3)

图中"α"称刚性角,按刚性角确定的基础,可安全地将荷载传给地基。对于砖基础一般 α=26°50′或33°50′,混凝土基础一般 α=45°。

2)非刚性基础,通常即钢筋混凝土基础(图2-4)

钢筋混凝土的翼部可以像悬臂板那样工作,不受刚性角控制,故基底 B 可以做得很宽,而高度 H 比刚性基础的高 H 却要小得多。

(3)根据剖面形式,基础有矩形、台阶形、锥形等。

2.2 Foundations

Definition: The part of a building below the surface of the ground is called the foundation of the building.

2.2.1 Foundation types

(1) According to the bearing area at ground base and the foundation geometric shapes(see Fig. 2-2)

1) Isolated or independent footing;
2) Continuous foundation or stripe footing including two-way continuous footing;
3) Raft foundation or mat foundation;
4) Reinforced concrete box foundation.

(2) According to foundation mechanic character

1) Rigid foundation(Fig.2-3)

In the figure, "α" is termed rigid angle, based on the rigid angle design the foundation which can safely pass loads to the ground base. For brick foundation, generally, α = 26°50′ or 33°50′, concrete foundation, α = 45°;

2) Non-rigid foundation is usually reinforced concrete (RC) foundation (Fig.2-4)

The alar parts of RC foundation can work as the cantilever slab does without the control of rigid angle, so the fundus B can be wider while the height H more shorter than that of the rigid foundation.

(3) According to the section shape, there are rectangle footing, stepped footing, taper footing, etc..

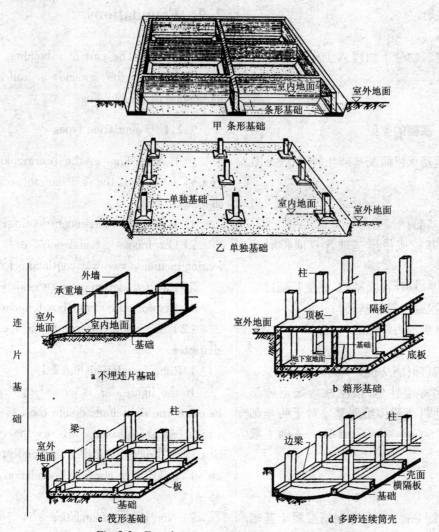

Fig.2-2 Foundation types according to the bearing area at ground base and foundation geometric shapes

图 2-2 根据承压面积与几何形状决定的基础类型

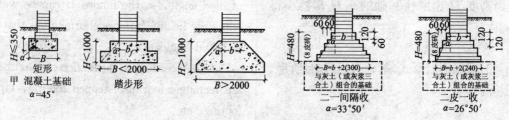

Fig.2-3 Rigid foundation

图 2-3 刚性基础例

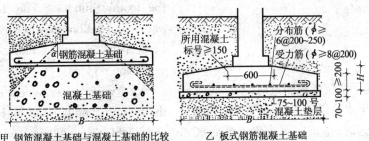

甲 钢筋混凝土基础与混凝土基础的比较　　乙 板式钢筋混凝土基础

Fig.2-4 An example:Non-rigid foundation-Rc foundation

图 2-4 非刚性基础例—钢筋混凝土基础例

注(Note):刚性基础 Rigid Foundations;非刚性基础 Non Rigid Foundations;混凝土基础 Concrete Foundations;钢筋混凝土基础 Reinforced Concrete(RC) Foundations;砖基础 Brick Foundations;连续基础 Continuous Footings;带(条)形基础 Tripe Footings;双向带形基础 Two-way Tripe Footings;单独基础 Isolated or Independent footings;钢筋混凝土箱形基础 RC Box Foundations;筏式基础 Raft or Mat Foundations。

2.2.2 基础最小埋置深度

影响因素及处理

(1)地基条件及荷载

1)软土 2~5m 下是好土:轻房基础可在软土层;重房基础落在好土层;

2)软土超过 5m 下是好土:轻房基础在软土层;对于重房可用换土或打桩处理;

3)软硬土交错土层:轻房基础可在某一选好的软土层或硬土层;重房应先换土或打桩处理。

(2)地下水

地下水是变动的,当地下水位上升到基础底面以上,地下水的浮力将导致地基承受的压力减小,反之,地基承受的压力又要增大到地下水未超过基础底面以前一样大。这样的变动易引起房屋不均匀下沉(特别是在湿陷区或湿胀区)。

2.2.2 Minimum buried depth of foundation

Influences and treatments

(1) Geologic conditions and loads

1) Hardpan beneath the soft soil of 2~5m: light house foundation may be at the soft soil; heavy house foundation must be at the hardpan;

2) Hardpan beneath soft soil of more than 5m: light house foundation may be at the soft; for heavy house may have soil changed or piled;

3) Alternatively soft soil and hardpan: light house foundation may be at a selected soft soil or hardpan; for heavy house, first the soil must be treated with soil changed or piled.

(2) Ground water

Ground water is fluctuant. When the water table is rising over the footing fundus, the ground water flotage(floatage) will cause abatement of the ground base pressure; contrarily, the pressure will increase to the same level as before as the water table below

因此，当地下水位离地面近时，基础底面应置于最低地下水位以下200mm左右；当地下水位离地面远时，基础底面应置于最高水位以上200mm左右，这样就可避免地下水变动的影响。

(3) 冻结

寒冷地区冷季，地面下的土壤会冻结到一定的深度，例如西安冻结深度为0.45m左右，内蒙有的地区冻结深度可达3m以上。基础底面若处在冻结深度内，冷季会受到冻胀上抬的作用，暖季消融期，该建筑又将回落。这种冻融循环，也易引起房屋不均匀下沉。

有两种方法避免这种冻融循环的影响：

传统方法是将基础底面置于冻结深度以下200mm左右；

另一方法是：对于强冻土如黏土、粉砂等，基础底面就放在冻结线上，其他冻土、弱冻土，基础底面可在冻结深度内高于冻结线一个d值（允许残留冻结深度）即可。因为在d值深度，冻融影响已很弱。这一方法比传统方法经济、合理（图2-5）。

the footing fundus. This fluctuation easily causes foundation unequal settlement(sinking) (especially in wet sinking region or wet expansion region).

So, when water table is near the ground, the footing fundus must be below the lowest water table some 200mm; when water table is far below the ground the footing fundus must be up over the highest water table some 200mm. If done so, the ground water influence will be avoided.

(3) Freeze

In cold region cold seasons, the undersoil will be freezed to a certain depth, for examples, in Xi'an region, the frost penetration is about 0.45m; in some places of Inner Mongolia Autonomous Region, the frost penetration is more than 3 m. If the footing fundus is within the frost penetration, during cold season it may be upward by the frost heave effect and downward during warm thaw season. This freezing-thawing circulation also easily causes unequal sinking (subsidence, subsiding) of the foundation.

There are two methods to avoid the effect of the freezing-thawing circulation:

Conventional method is to set the footing fundus below the frost penetration some 200mm;

Another method: for strong freezed soil such as clay, powdery sand… set the footing fundus at the frost line; for other freezed soil or weakly freezed soil, set the footing fundus within the frost penetration with a "d" mm higher than the frost line("d" is the allowable residual frost penetration). Because within the allowable frost penetration, the freezing-thawing effect is very weak. This method is more economical and rational than

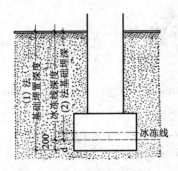

Fig2-5 A comparison between the two methods deciding footing depth
图 2-5 两种方法基础埋深比较

对于岩石、卵石、砾石、粗砂等,冻融循环影响很小,在这些地基上建房可不考虑冻融影响。永冻区,无冻融影响。

(4)地下室

当建筑物有地下室时,地下室外墙与该墙基础墙结合为一体。基础常采用浮筏式基础(满堂基础)或箱形基础。该外墙既是挡土墙也是挡水墙,对整个地下室必须做好防水处理。

(5)相邻建筑

当靠近旧房建新房时应按如下处理:

①两基础处于同一深度水平;

②$L \geqslant 2\Delta H$。

L:两基础间水平净距离;

ΔH:两基础底面之间垂直距离。(图 2-6)

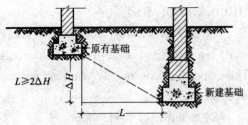

Fig.2-6 The relation between an old foundation and a new one
图 2-6 新老基础关系

(6)构造要求

基础最小埋置深度 $D \geqslant 500mm$

conventional one.

For rock, pebble, gravel, coarse sand (torpedo sand, harsh sand, grit), etc. the freezing-thawing circulation effect is very weak. To build on these ground bases, we do not consider the freezing-thawing effect. In permafrost region, no freezing-thawing effect.

(4) Basement

When a building with a basement whose exterior wall is combined foundation wall together. Draft (mat) foundation or box foundation is usually used. The exterior wall is not only a retaining wall but also a water-resisting wall, for the whole basement waterproof work must be treated well.

(5) Adjacent buildings

When nearby an old building to build a new one, should do as follows:

① Both the two foundations at the same depth level;

② $L \geqslant 2\Delta H$。

L: the horizontal net distance between the two foundations;

ΔH: the vertical distance between the two foundation bottoms.

(6) Construction demand

Foundation minimum buried depth: $D_{min} \geqslant 500mm$

2.3 墙

思考:钢、铁、铝、木、土、砖、石、竹、混凝土、冰、石灰、石膏、纸、塑料、玻璃、空气,它们当中哪些可作为墙的材料? 哪些不能作为墙的材料?

2.3.1 黏土砖墙(简称砖墙)

(1)材料与砌砖

砖墙由砖与砂浆砌筑而成。

制砖过程:取土→过筛→和料→成型→焙烧→产品输出

黏土砖的强度等级:

MU30,MU25,MU20,MU15,MU10,MU7.5(MPa)。

$1MPa = 10.197 kgf/cm^2$

砌墙砂浆常用水泥:石灰膏:砂按一定体积比或重量比混和后加水拌制而成,通称混合砂浆。砂浆强度等级:M0.4、M1.0、M2.5、M5.0、M7.5、M10.0、M15.0(MPa)。

图 2-7 表示出了实心黏土砖的规格及常用几种砌法。该规格主要考虑了手工操作方便而定。

(2)砖墙的稳定性

增强砖墙稳定性通常有两种方法:

1)设圈梁,即在外墙和选定的内纵、横墙各层楼层标高处设置高 240~360mm,宽同墙宽的钢筋混凝土圈梁,在地震区,一般每层楼层处均设,并和构造柱联结在一起组成

2.3 Walls

Thinking deep: steel, iron, aluminium, wood(timber), soil, brick, stone, bamboo, concrete, ice, lime, gypsum, paper, plastic, glass, water and air, which of them can be wall materials? Which of them can't be?

2.3.1 Clay brick walls(brick walls)

(1) Material and bricklaying

Brick wall is laid by bricks and mortar. Bricks are made by the following processes:

Soil dug →sifted→admixed→modeled→baked→products out

Intensity grades of clay brick:

MU30, MU25, MU20, MU15, MU10, MU7.5(MPa)

$1MPa = 1millon(mege)Pa = 10.197 kgf/cm^2$

Bricklaying mortar is made of cement, hydrated lime and sand, they are in a set proportioning on volume or weight and mixed by a set water, it's usually called mixed mortar.

Intensity grades of mortar:

M0.4, M1.0, M2.5, M5.0, M7.5, M10.0, M15.0(MPa)

Fig.2-7 is showing the specification of solid clay brick and some usual bricklayings. These specifications mainly consider for easy artificial bricklaying.

(2) Stability of brick walls

Generally there are two ways to strengthen brickwall stability:

1) Set girt(girth), at each floor level of the exterior wall and selected inner lengthways (lengthwise) walls and cross walls set a RC girt of 240mm~360mm height with a width

增强稳定性和刚度的钢筋混凝土框架(图2-8a);2)设壁柱(图2-8b)。

as the same of the wall. In earthquake region, generally, each floor level, one circular girt. Girts and construction posts combine to form a RC fram to strengthen wall stability and rigid (rigidity)(Fig.2-8a);2)set pilasters(Fig.2-8b).

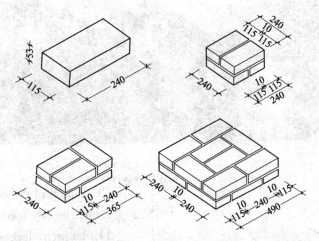

Fig. 2-7 Solid clay brick specifications and normal bonds

图 2-7 实心黏土砖规格及常用砌法

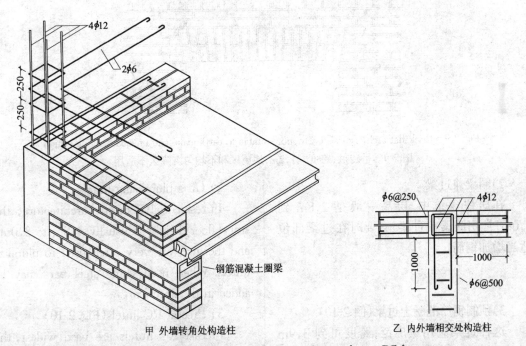

甲 外墙转角处构造柱　　　　　乙 内外墙相交处构造柱

Fig.2-8a) Girts and construction posts combine to form a RC fram to strengthen wall stability and rigid(rigidity)

图 2-8a 圈梁与构造柱结合成 RC 框架增强砖墙稳定性与刚度

Fig.2-8b) Set pilasters
图 2-8b 设壁柱

(3)门窗过梁
常用过梁：
1)平拱(图 2-9)

(3) Lintels of doors and windows
Lintels usually used：
1) Flat arch, jack arch or straight arch. (Fig.2-9)

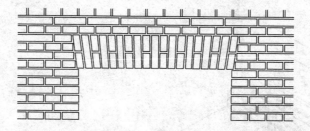

Fig.2-9 Brick flat arch, span≤1.2m, not suitable to earthquake and unequal sinking region
图 2-9 平砖拱,跨度≤1.2m,地震区和不均匀沉降区不适用

2)圈梁兼过梁

在居住建筑中,层高一般是 2.8m 到 3.0m,常用圈梁兼过梁,必要时在过梁部位适当增加钢筋。

2) Girts pluralizing lintels

In residential(domestic) architecture, the story(floor) height is usually 2.8m to 3.0m, and the girts are very often used to pluralize lintels. When need, some steel bars may be added in the lintel part.

3)预制钢筋混凝土过梁(图 2-10)

这种过梁应用最广泛,跨度可到 3.9m 至 4.8m。

3) Precast RC lintels(Fig.2-10)

Precast RC lintels are used widest, the span may reach 3.9m to 4.8m.

4)砖拱或石拱过梁(图 2-11)

这种过梁常用在要求美学效果更高处。

4) Brick/stone arch lintels(Fig.2-11)

Brick/stone arch lintels are used in where

在建筑中,跨度可到 4.8m 至 6.0m。用于桥梁,跨度可更大。

needs higher aesthetic effect. In buildings, the span may reach 4.8m to 6.0m. In building bridges, the span can reach more wider.

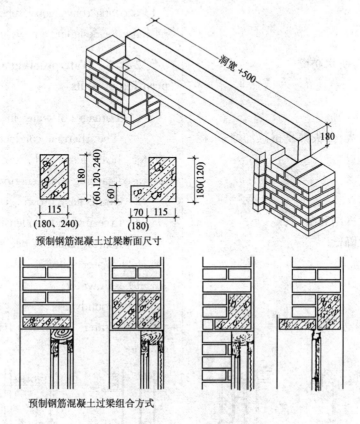

Fig.2-10　Precast RC lintels

图 2-10　装配式 RC 过梁

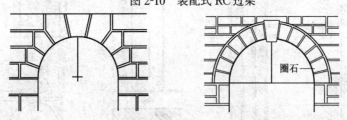

Fig.2-11　Brick/stone arch lintels

图 2-11　砖/石拱过梁

(4) 烟道、通风道、垃圾道

烟道、通风道从安全与卫生要求考虑,用户之间不宜串通。垃圾道应改造,以便其适合垃圾分类重力下达底层集装箱化的垃圾箱,由垃圾处理中心定时运走更换。图 2-12 为烟道及通风道,图 2-13 为传统垃圾道,设

(4) Smoke flue, air flue, rubbish flue

Based on safety and health, the smoke flues or air flues between users should not be serial with each other. Conventional rubbish flues should be reformed to fit for the categorized rubbish gravitating into the containerized container at the

计新式垃圾道可参考。

2.3.2 墙身防水防潮

(1)墙内含水的危害

1)导热率增大,冬失热夏得热均增多,耗能加大;

2)易生细菌;

3)可能产生冻结和凝结;

4)承重能力降低。

(2)水的来源

1)降雨;

2)地下水;

3)人为水:供水、排水。

bottom of the building, which will be carried away on time by Rubbish Center.

Fig. 2-12 shows smoke flue and air flue, Fig. 2-13 showing conventional rubbish flue can be a reference for designing a new one.

2.3.2 Waterproofing and damp-proofing of walls

(1)Damages of water in wall

1) The thermal conductivity increased, winter heat loss and summer heat gain increased causing more energy consumed;

2) Germs easy growing;

3) Freezing and condensing potential;

4) Bearing capacity decreased.

(2) Water sources

1) Rainwater;

2) Groundwater;

3) Water by man: water supply and drainage.

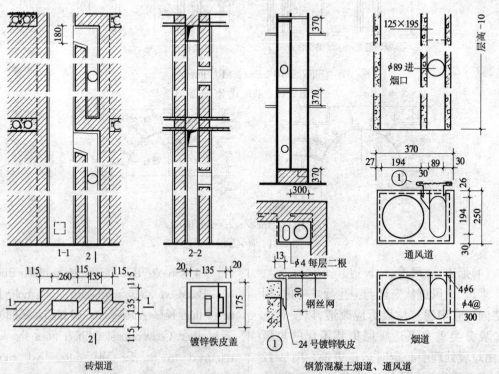

Fig.2-12 smoke flues, air flues　　图 2-12　烟道、通风道

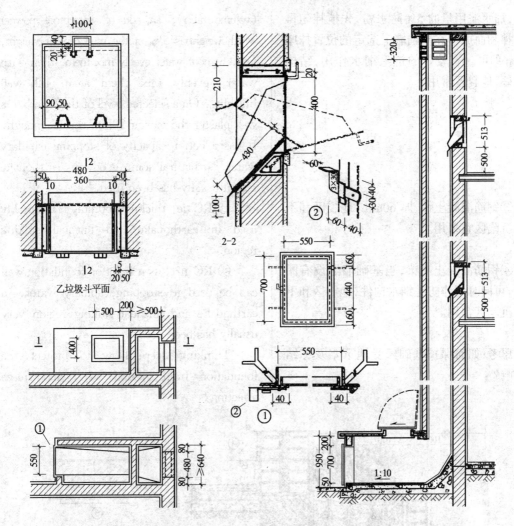

Fig.2-13 Showing the conventional rubbish flue may be a reference for the new design

图 2-13 图示为传统垃圾道可作为新设计的参考

(3)墙身防水防潮方法

图 2-14 为墙身(以砖墙为例)防水防潮构造。

1)水平防潮层

①一毡二油防潮层,由两层沥青胶夹一层油毡组成。抗毛细水渗透能力强,但与灰浆层联结不牢,故地震区不应采用。

②防水砂浆,1:2.5(水泥:砂子(体积

(3) Waterproof and damp-proof methods of walls

Fig 2-14 shows the construction of waterproof and damp-proof of a exterior wall.

1) Horizontal damp-proof course

① One layer of asphalt felt between two layers of mastic, it has high capacity resisting capillary water penetration, but weak join between it and the mortar. It should not be used in earthquake region.

② Waterproof mortar, cement: sand=1:2.5

比)),加水泥用量的5%防水粉,先搅拌匀再加水拌和而成。在基础墙上选定的位置抹厚20mm即可。有较好的抗毛细水作用,结构联结好,地震区常用。

③钢筋混凝土带,厚60mm,地震区和不均匀沉陷区宜采用。

④钢筋混凝土圈梁,当基础墙上设有圈梁时,可同时用作防毛细水构件,地震区和不均匀沉陷区常用。

思考:如何保护基础经受地下水及冻结的影响?

(volume ratio), an add of waterproof powder which weight is 5% of the used cement weight, first, without water evenly mix them, then with water, evenly mix them again till well finishing. On a selected level of the foundation wall plaster the mortar 20mm that is enough. It has a better capacity of stopping capillary water, structural joint good, in earthquake region usually has been used.

③ RC tie, thickness 60mm, it is suitably used in earthquake and unequal sinking region.

④ RC girt, as it is set on foundation wall can be used to stopping capillary water, in earthquake and unequal sinking region very usually has been used.

Thinking deep: How to protect the foundation from ground water and freeze affecting?

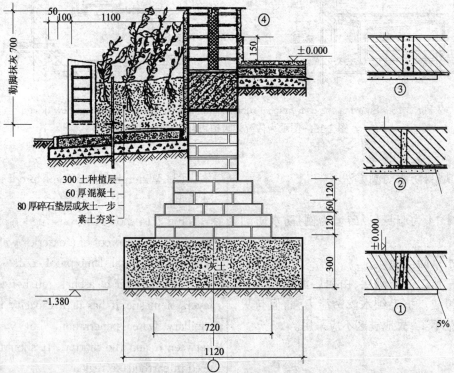

Fig. 2-14 Waterproofing and damp-proofing of wall and apron planting

图 2-14 墙身防水防潮及散水种植

2) 散水及其绿化

细部见图 2-14④。散水的传统功能是将墙面及屋面(自由落水或有管排水)的雨水排走,以便保护地基及基础。

散水宽度一般为 1.0m～1.2m。湿陷区,房高 8m 后,每增高 4m,散水增宽 250mm,最宽不得超过 2.5m。

思考:散水将雨水排走是一种浪费,你有何改进措施?

为防止混凝土散水干缩或冷缩产生的破坏性裂缝,该散水应设规格的分隔缝将散水分开,间距 4～6m,分隔缝宽 20mm,用弹性防水材料如沥青石膏粉填充。

散水直角相交的阳角或阴角处以及散水与勒脚抹灰相交处均应设分隔缝并严密填以弹性防水材料。

图 2.14④所示在传统散水上设 25cm 土种植层,即可种植花卉及爬山虎等,其优点是:①利用太阳能与雨水;②增强基础冬夏绝热;③消除回溅水;④绿化生态正效应;⑤变传统散水—非可持续构件为可持续构件。

3) 勒脚抹灰(图 2-14④)

为防止天空降雨及檐口水的回溅水侵蚀墙体,应做勒脚抹灰,水泥:砂子 1:2.5 体积比灰浆,厚 20mm,高 700mm 即可(经实际观察与试验证实:该回溅水高度超不过 650mm,也

2) Apron and its greening

The detail is showing in Fig. 2-14④. The conventional function of apron is to drain off the rainwater from wall and roofing (by free-fall or piping), so as to protect ground base and foundation.

The ordinary width of apron is 1.0m～1.2m. In unequal sinking region, when the building height over 8m, as every increasing height of 4m, the apron width should be 250mm increased, but the maxmum width should not be over 2.5m.

Thinking deep: Apron drains away rainwater that is a waste, do you have any improvement?

To forfend the destructive cracks on concrete apron caused by dry or cold contractions, the apron should be separated by the standard gaps of spacing 4～6m, each gap width 20mm filled by elastic waterproof material such as asphalt (bitumen)-gypsum powder.

The apron orthogonal outside corners and inside ones and the joints between the apron and plinth finish should also be set the gaps and tightly filled with elastic waterproof material.

Fig 2.14④ shows that on the conventional apron set a 25cm soil planting layer we may plant flowers or Boston ivy, etc. which benefits: ①using solar energy and rainwater; ②increasing winter and summer insulation; ③cancelling rebounding water; ④ green positive eco-effect; ⑤making the conventional apron non-sustainable element sustainable one.

3) Plinth finish (Fig.2-14④)

To forfend the plinth wall imbued by the rebounding water of sky rain and eaves water a plinth finish should be set with cement:sand mortar of a volume ratio 1:2.5, thickness

不存在房愈高回溅水也愈高的现象）。

4）凡凸出外墙的构件如窗台、挑檐、阳台以及过梁、女儿墙压顶等均宜做尖劈式滴水，试验证实比传统槽式或凸缘式滴水好，尖劈式滴水可阻止大、小雨水绕溅墙面，槽式或凸缘式滴水阻止不了大雨绕过该滴水淋湿墙面。(图2-15)。

20mm, height 700mm that is enough. Practical observing and experiments have attested that the rebounding water never over a height of 650mm, and there is no any phenomenon of "the taller building, the higher rebounding water".

4) All the protruded elements on exterior wall such as windowsills, overhangs, balconies and lintels and parapet coping, etc. suit to have wedged drips that by experimental attestation are better than the conventional groovy or flange drips. Wedged drip can stop either light or heavy rainwater circumanbulating to imbue the wall, but groovy or flange drip can't stop the heavy rainwater to imbue the wall.(Fig.2-15)

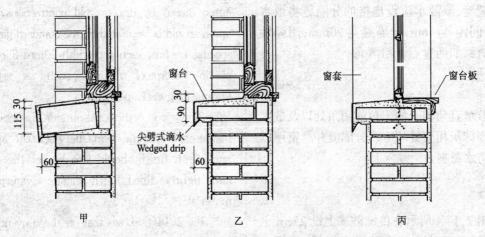

Fig.2-15 Drip construction　　图2-15 滴水构造

5）潮湿房间如浴室、盥洗室、厕所、厨房等墙面常贴瓷砖，地面常铺锦砖（马赛克）防水防潮并做地漏将污水排入下水管。

思考：内墙要做水平防潮层吗？

5) In wet rooms such as bathroom, washroom, lavatory and kitchen (cookroom), etc. the walls are often adhibited with glazed tiles and floor with mosaic provement. And the sewage is drained by drainpipe through scupper.

2.4 冬季建筑热学

2.4 Building Heat in Winter

思考：1支40瓦的灯冬季能保持室内

Thinking deep: Can a 40W lamp keep

20℃吗？

（1）启思图与启思式

$$Q = \frac{t_i - t_o}{R_0} F \cdot Z \quad (J) \quad (2.3.2\text{-}1)$$

图2-16 为冬季外墙热工动态，是一启思图。公式(2.3.2-1)是一启思式。

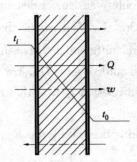

Fig. 2-16 The thermal dynamic showing of an exterior wall in winter

图2-16 冬季外墙热工动态图

图中和式中：

$t_i \sim t_0$ 曲线为室内→墙剖面→室外温度降落曲线；

ω：水蒸气渗透；

←⋯：新鲜空气渗入；

⋯→：脏空气渗出；

Q：失热量同时也是供热量(J,焦耳)；

t_i：室内气温℃；

t_0：室外气温℃；

F：失热面积 m^2；

Z：失热时间 s；

R_0：外墙总热阻 $m^2 \cdot K/W$，$1W = 1J/s$，

(1瓦=1焦耳/秒)，故 $m^2 \cdot K/W = m^2 \cdot K \cdot s/J$。

例：

某墙 $R_0 = 0.5\ m^2 \cdot K \cdot s/J = m^2 \cdot K/W$，表明当室内外温差1K，通过 $1m^2$ 墙面，由室内向室外失热1焦耳需要0.5秒的时间。

indoor 20℃ in winter?

（1） An enlightening picture and an enlightening formula

Fig. 2-16, the thermal dynamic showing of an exterior wall in winter is an enlightening picture.

Formula (2.3.2-1) is an enlightening formula.

In the picture and formula：

$t_i \sim t_0$: is the temperature drop curve from indoor→wall section→outdoor；

ω: water vapour (aqueous vapour) penetration；

←⋯: fresh air in；

⋯→: dirty air out；

Q: heat loss also heat supply, (J, joul)；

t_i: indoor temperature, ℃；

t_0: outdoor temperature, ℃；

F: heat loss area, m^2；

Z: heat loss time, s；

R_0: exterior wall total thermal resistance, $(m^2 \cdot K)/W$, $1W = 1J/s$, $\therefore m^2 \cdot K/W = m^2 \cdot K \cdot s/J$.

Example:

A wall $R_0 = 0.5\ m^2 K \cdot s/J$, it explains that when the indoor-outdoor air temperature differential is 1K, through $1m^2$ wall area from indoor to outdoor, heat loss 1J needs the time of 0.5 second.

$R_0 = R_i + R_W + R_e$

R_i:感热阻,一般 $R_i = 0.115 \text{m}^2 \cdot \text{K/W}$,即当室内与外墙内表面温差1K,$1\text{m}^2$ 墙面感受1J热量需要0.115s(秒)时间。

R_e:放热阻,一般 $R_e = 0.043 \text{m}^2 \cdot \text{K/W}$,即当外墙外表面与室外温差1K,$1\text{m}^2$ 墙面放出1J热需要0.043s时间。

思考:为什么 R_e 比 R_i 小很多?

$R_W = d/\lambda \quad \text{m}^2 \cdot \text{K/W}$

d:墙厚 m;

λ:导热系数(热导率)W/(m·K),例如某黏土砖砌体 $\lambda = 0.81 \text{W/(m·K)}$,即该砖砌体厚1m,两表面温差1K,在一维传热条件下高温面每秒传给低温面0.81J热量。一般材料热工指标见表2-2。

$R_0 = R_i + R_W + R_e$

R_i: inner surface resistance, general $R_i = 0.115 \text{m}^2 \text{K/W}$, it explains that when the temperature differential between indoor air and the inner surface of the exterior wall is 1K, 1 m^2 of the surface accepts 1J heat that needs 0.115 second time;

R_e: outer surface resistance, general $R_e = 0.043 \text{m}^2 \cdot \text{K/W}$, it means that when the temperature differential between the outer surface and outdoor air is 1K, 1m^2 of the outer surface area releases 1J heat that needs 0.043 second time;

Thinking deep: Why the R_e is so smaller than R_i?

$R_W = d/\lambda \quad \text{m}^2 \cdot \text{K/W}$

d: thickness of the wall m;

λ: thermal conductivity, W/(m·K), Example: A clay brickwork $\lambda = 0.81 \text{W/(m·K)}$, i.e. the brickwork of 1m thickness, between the two surfaces, the temperature differential is 1K, in one direction heat transmission, from warm surface to the cool surface in one second can transfer 0.81 joule heat.

General material thermal data see table 2-2

表2-2 建筑材料热工指标

材料名称	容重 γ 〔kg/m³〕	导热系数 λ 〔W/(m·K)〕	蓄热系数 S_{24} 〔W/(m²·K)〕	比热 c 〔kJ/(kg·K)〕	蒸汽渗透系数 $\mu \times 10^4$ 〔g/(m·h·Pal)〕
一、混凝土					
钢筋混凝土	2500	1.74	17.20	0.92	0.158
碎石、卵石混凝土	2300	1.51	15.36	0.92	0.173
碎石、卵石混凝土	2100	1.28	13.50	0.92	0.173
膨胀矿渣珠混凝土	2000	0.77	10.54	0.96	—
膨胀矿渣珠混凝土	1800	0.63	9.05	0.96	0.975
膨胀矿渣珠混凝土	1600	0.53	7.87	0.96	1.05
自然煤矸石、炉渣混凝土	1700	1.00	11.68	1.05	0.548

续表

材料名称	容重 γ [kg/m³]	导热系数 λ [W/(m·K)]	蓄热系数 S_{24} [W/(m²·K)]	比热 c [kJ/(kg·K)]	蒸汽渗透系数 $\mu \times 10^4$ [g/(m·h·Pa)]
自然煤矸石、炉渣混凝土	1500	0.76	9.54	1.05	0.900
自然煤矸石、炉渣混凝土	1300	0.56	7.63	1.05	1.05
粉煤灰陶粒混凝土	1700	0.95	11.40	1.05	0.188
粉煤灰陶粒混凝土	1500	0.70	9.16	1.05	0.975
粉煤灰陶粒混凝土	1300	0.57	7.78	1.05	1.05
粉煤灰陶粒混凝土	1100	0.44	6.30	1.05	1.35
黏石陶粒混凝土	1600	0.84	10.36	1.05	0.315
黏石陶粒混凝土	1400	0.70	8.93	1.05	0.390
黏石陶粒混凝土	1200	0.53	7.25	1.05	0.405
页岩陶粒混凝土	1500	0.77	9.70	1.05	0.315
页岩陶粒混凝土	1300	0.63	8.16	1.01	0.390
页岩陶粒混凝土	1100	0.50	6.70	1.05	0.435
浮石混凝土	1500	0.67	9.09	1.05	—
浮石混凝土	1300	0.53	7.54	1.05	0.188
浮石混凝土	1100	0.42	6.13	1.05	0.353
加气、泡沫混凝土	700	0.22	3.56	1.05	1.54
加气、泡沫混凝土	500	0.19	2.76	1.05	1.99
二、砂浆和砌体					
水泥砂浆	1800	0.93	11.26	1.05	0.900
石灰、水泥复合砂浆	1700	0.87	10.79	1.05	0.975
石灰砂浆		0.81	10.12	1.05	1.20
石灰、石膏砂浆	1500	0.76	9.44	1.05	—
保温砂浆	800	0.29	4.44	1.05	—
重砂浆砌筑黏土砖砌体	1800	0.81	10.53	1.05	1.05
轻砂浆砌筑黏土砖砌体	1700	0.76	9.86	1.05	1.20
灰砂砖砌体	1900	1.10	12.72	1.05	1.05
重砂浆砌筑 26、33 及 36 孔黏土空心砖砌体	14900	0.58	7.52	1.05	1.58
三、热绝缘材料					
矿棉、岩棉玻璃棉 板	<150	0.064	0.93	1.22	4.88
矿棉、岩棉玻璃棉 板	150~300	0.07~0.093	0.98~1.60	1.22	4.88
矿棉、岩棉玻璃棉 毡	≤150	0.058	0.94	1.34	4.88
矿棉、岩棉玻璃棉 松散	≤100	0.047	0.56	0.84	4.88
膨胀珍珠岩、蛭石制品					
水泥膨胀珍珠岩	800	0.26	4.16	1.17	0.42
	600	0.21	3.26	1.17	0.90
	400	0.16	2.35	1617	1.91
沥青、乳化沥青膨胀珍珠岩	400	0.12	2.28	1.55	0.293
	300	0.093	1.77	1.55	0.675
水泥膨胀蛭石	350	0.14	1.92	1.05	
泡沫材料及多孔聚合物					
聚乙烯泡沫塑料	100	0.047	0.69	1.38	
	30	0.042	0.35	1.38	0.144
聚乙烯泡沫塑料	50	0.037	0.43	1.38	0.148
	40	0.033	0.36	1.38	0.112

续表

材料名称	容重 γ [kg/m³]	导热系数 λ [W/(m·K)]	蓄热系数 S_{24} [W/(m²·K)]	比热 c [kJ/(kg·K)]	蒸汽渗透系数 $\mu \times 10^4$ [g/(m·h·Pal)]
四、建筑板材					
胶合板	600	0.17	4.36	2.51	0.225
软木板	300	0.093	1.95	1.89	0.255
	150	0.058	1.09	1.80	0.285
纤维板	600	0.23	5.04	2.51	1.13
石棉水泥板	1800	0.52	8.57	1.05	0.135
石棉水泥隔热板	500	0.16	2.48	1.05	3.9
石膏板	1050	0.33	5.08	1.05	0.79
水泥刨花板	1000	0.34	7.00	2.01	0.24
	700	0.19	4.35	2.01	1.05
稻草板	300	0.105	1.95	1.68	3.00
木屑板	200	0.065	1.41	2.10	2.63
五、松散材料					
无机材料					
锅炉渣	1000	0.92	4.40	0.92	1.93
高炉炉渣	900	0.26	3.92	0.92	2.03
浮石	600	0.23	3.05	0.92	2.63
膨胀珍珠岩	120	0.07	0.84	1.17	1.50
	80	0.058	0.63	1.17	1.50
有机材料					
木屑	250	0.093	1.84	2.01	2.63
稻壳	120	0.06	1.02	2.01	
石油沥青	1400	0.27	6.73	1680	
	1050	0.17	4.71	1680	0.075
平板玻璃	2500	0.76	10.69	840	0
玻璃钢	1800	0.52	9.25	1260	
建筑钢材	7850	58.2	126.1	480	0

注(Note)：砾石、卵石混凝土 Gravel or Pebble Concrete；膨胀矿渣混凝土 Expanded slag concrete；自然煤矸石、炉渣混凝土 Natural Gangue Concrete, Cinder Concrete；粉煤灰陶粒混凝土 Coal Powder Ceramsite Concrete；黏土陶粒混凝土 Clay Rock Seramsite Concrete；加气混凝土 Aerocrete；泡沫混凝土 Foam Concrete；水泥砂浆 Cement Sand Mortar；石灰、水泥复合砂浆 Lime Cement & Sand Mixed Mortar；石灰砂浆 Lime-Sand Mortar；石灰、石膏砂浆 Lime Gypsum & Sand Mixed Mortar；保温砂浆 Insulation Mortar；重砂浆砌筑黏土砖砌体 Clay Brick Masonry by Heavy Mortar；轻砂浆砌筑黏土砖砌体 Clay Brick Masonry by Light Mortar；灰砂砖砌体 Cement-Sand Brick Masonry；重砂浆砌筑黏土空心砖砌体 Hollow Clay Brick Masonry by Heavy Mortar；矿棉板 Mineral Wool Board；岩棉板 Rock Wool Board；Glass Wooll Board；水泥膨胀珍珠岩 Cement Expanded Pearlite；沥青膨胀珍珠岩 Asphalt Expanded Pearlite；乳化沥青膨胀珍珠岩 Emulsified Asphalt Expanded Pearlite；水泥膨胀蛭石 Cement Expanded Vercumlite；聚乙烯泡沫塑料 Foamed Polyethylene；胶合板 Plywood；软木板 Coak Board；纤维板 Fibre Board；石棉水泥板 Asbestos Cement Board；石棉水泥隔热板 Asbestos Cement Insulation Board；石膏板 Gypsum Board；水泥刨花板 Cement Shaving Board；稻草板 Rice-Straw Board；木屑板 Wood Sorape Board；锅炉渣 Boiler Slag(cinder)；高炉炉渣 Blast-Furnace Slag(Cinder)；浮石 Pumice；膨胀珍珠岩 Expanded Pearlite；木屑 Wood Scrape；稻壳 Rice Husk(Hull)；石油沥青 Petroleum Pitch；平板玻璃 Plate Glass；玻璃钢 Glass Fibre Reinforced Plastic；建筑钢材 Building Steel Materials

续表 2-2

标准大气压时不同温度下的饱和水蒸气分压力 P_s 值(Pa)
a. 温度自 0℃ 至 −20℃（与冰面接触）

t(℃)	0.0	0.1	0.2	0.3	0.4	0.5	0.6	0.7	0.8	0.9
−0	610.6	605.3	601.3	595.9	590.6	586.6	581.3	576.0	572.0	566.6
−1	562.6	557.3	553.3	548.0	544.0	540.0	534.6	530.6	526.6	521.3
−2	517.3	513.3	509.3	504.0	500.0	496.0	492.0	488.0	484.0	484.0
−3	476.0	472.0	468.0	464.0	460.0	456.0	452.0	448.0	445.3	441.3
−4	437.3	433.3	429.3	426.6	422.6	418.6	416.0	412.0	408.0	405.3
−5	401.3	398.6	394.6	392.0	388.0	385.3	381.3	378.6	374.6	372.0
−6	368.0	365.3	362.6	358.6	356.0	353.3	349.3	346.6	344.0	341.3
−7	337.3	334.6	332.0	329.3	326.6	324.0	321.3	318.6	314.7	312.0
−8	309.3	306.6	304.0	301.3	298.6	296.0	293.3	292.0	289.3	286.6
−9	284.0	281.3	278.6	276.0	273.3	272.0	269.3	266.6	264.0	262.6
−10	260.0	257.3	254.6	253.3	250.6	248.0	246.6	244.0	241.3	240.0
−11	237.3	236.0	233.3	232.0	229.3	226.6	225.3	222.6	221.3	218.6
−12	217.3	216.0	213.3	212.0	209.3	208.0	205.3	204.0	202.6	200.0
−13	198.6	197.3	194.7	193.3	192.0	189.3	188.0	186.7	184.0	182.7
−14	181.3	180.0	177.3	176.0	174.7	173.3	172.0	169.3	168.0	166.7
−15	165.3	164.0	162.7	161.3	160.0	157.3	156.0	154.7	153.3	152.0
−16	150.7	149.3	148.0	146.7	145.3	144.0	142.7	141.3	140.0	138.7
−17	137.3	136.0	134.7	133.3	132.0	130.7	129.3	128.0	126.7	126.0
−18	125.3	124.0	122.7	121.3	120.0	118.7	117.3	116.6	116.0	114.7
−19	113.3	112.0	111.3	110.7	109.3	108.0	106.7	106.0	105.3	104.0
−20	102.7	102.0	101.3	100.0	99.3	98.7	97.3	96.0	95.3	94.7

影响"λ"的主要因素：

① 密度 ρ：实际常按容重测出"λ"值，标名仍为 ρ，如砖砌体 $\rho = 1800 kg/m^3$，$\lambda = 0.81 W/(m·K)$；钢 $\rho = 7850 kg/m^3$，$\lambda = 58.2 W/(m·K)$，此处 ρ 实际是材料的容重。一般说密度或容重愈大，热导率也愈大，但也有例外（见(3)和(5)）。

② 材料湿度：材料内部空气被水侵入，湿度就加大。一般空气的 $\lambda = 0.029$，而水的

Main factors affecting "λ":

① Density ρ: actually the "λ" is measured by its unit weight (also marked ρ), such as brickwork $\rho = 1800 kg/m^3$, $\lambda = 0.81 W/(m·K)$; steel $\rho = 7850 kg/m^3$, $\lambda = 58.2 W/(m·K)$, here ρ actually is material's unit weight. In general, the more density or unit weight the more thermal conductivity but exceptions (see (3) and (5)).

② Humidity: the air pores in material occupied by water, the humidity of the

λ=0.58,是一般空气 λ 值的 20 倍,故材料的湿度增大,λ 值也必增大。

③材料内部微孔结构:内部微孔密闭互不相通的材料与内部微孔相通形成较大空隙的材料比较,前者容重较重但热导率却较小,因为后者有利于对流换热;但后者吸声效果较前者好,因为后者有利于空气声波通过摩擦将声能转化成热能。

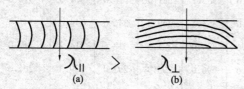

Fig.2-17 Relation between λ and heat flow direction
图 2-17 热导率与热流方向的关系

④热流方向:图 2-17(a)热流平行于木纹(年轮)传递,(b)热流垂直于木纹传递,热导率后者小于前者($\lambda_\perp < \lambda_\parallel$)。

⑤材料分子、原子、电子参与热运动的活跃性:钢密度 7850kg/m³ 是铝密度 2600kg/m³ 的三倍多,但钢的热导率 58.2W/(m·K)却是铝的热导率 190W/(m·K)的 1/3 还小。这是铝材的分子、原子、电子参与热运动的活跃性更大的缘故。

⑥时间:有的材料,时间长了,热导率会变大。采用预制绝热材加以塑封或腊封是避免时间对热导率影响的好办法。

从式(2.3.2-1)可知:
当一座建筑物例如一幢住宅在某地建成,则其室外温度 t_0、失热面积 F、失热时间亦即供热时间 Z 均为定值,那么,如果有足

material increased, common air λ = 0.029, water λ = 0.58, 20 times that of air. So the more humidity of material the more thermal conductivity of them.

③ Material inner porous composition: material with pores isolated each other comparing to another with communicated pores, the unit weight of the former is larger than that of the latter, but the thermal conductivity smaller than that of the latter due to in the latter easy convective heat transmission, and yet the sound absorptance of the latter is better than that of the former, because the latter benefits the air sound waves to convert the sound energy into heat energy through friction.

④ Heat flow direction: In Fig. 2-17(a) heat flow is paralleled to the wood grains (annual rings), in (b) heat flow vertical to the wood grains, the thermal conductivity of the latter is smaller than that of the former($\lambda_\perp < \lambda_\parallel$).

⑤ The activity of material's molecule, atoms and electrons in thermal moving: steel's unit weight 7850kg/m³ is more than three times aluminium unit weight 2600kg/m³, but the former's thermal conductivity 58.2W/(m·K) is less than one third of the latter's thermal conductivity 190 W/(m·K) that is due to more active in thermal moving of the moleculae, atoms and electrons of aluminium.

⑥ Time: In some materials, after long time the thermal conductivities become larger. Prefab insulation with plastic or paraffine envelop can minimize the time effect.

From formula(2.3.2-1)we can see:
When a building such as a dwelling has been built, the outdoor air temperature t_0, heat loss area F and heat loss time i.e. heat supply time Z all have become constant now, if there is an

够大的总热阻 R_0，就可用很少的热量 Q 维持室内某一定值的空气温度例如 20℃。40W 的灯除发光外必然有一恒量的余热存在。因此，对思考题的回答是：如果有足够的总热阻 R_0，40W 的灯冬季可保持室内 20℃。但这是理论的可以，实际上并未做到，这就意味着要用很少的供热量冬季保持室内一定的温度（节能建筑）还有很多科研工作要做，提高 R_0 就是其中之一。

(2) 必需热阻与实有热阻

围护结构的实有热阻必须满足必需热阻的要求。

20 世纪 50 年代至 80 年代，我国建筑围护结构的最低必需总热阻按下式确定：

$$R_{0\cdot\min\cdot N} = \frac{t_i - t_o}{[\Delta t]} R_i \quad \text{m}^2 \cdot \text{K/W}$$

(2.3.3 - 2)

$R_{0\cdot\min\cdot N}$：最低必需总热阻 $\text{m}^2\text{K/W}$;

t_i：室内必需空气温度，℃；

t_o：室外计算空气温度，℃；

R_i：感热阻 $\text{m}^2 \cdot \text{K/W}$;

$[\Delta t]$：容许的室内气温与围护结构内表面温度之差，℃ 或者 K，表 2-3 为例。

表 2-3　室内气温与外围护结构内表面容许温差值

$[\Delta t]$(K)	外墙	屋顶
居建、医疗、托幼	6	4
办公、学校、门诊	6	4.5
公建（除上述外）	7	5.5

[例题]　确定西安某住宅砖外墙厚度。设 $t_i = 18℃, t_o = -8℃$

enough big total thermal resistance R_0 we could keep indoor air temperature at a certain degree such as 20℃ with only a little heat Q. When a 40W lamp is lighting there certainly a constant remaining heat existing. So, to answer the cogitative question we can say if there is an enough big total thermal resistance R_0, the 40 W lamp could keep indoor at 20℃, that is only a theoretic "might", in fact no one has done. It means that with a little heat to keep indoor a certain temperature (energy saving building) a lot of works waiting us to do, to increase R_0 is one of the works.

(2) The needful thermal resistance and existing thermal resistance

The envelop's existing thermal resistance should satisfy the needful thermal resistance.

During 1950s ~ 1980s in China the minimum needful total thermal resistance of envelops is quantified by the following formula:

$$R_{0\cdot\min\cdot N} = \frac{t_i - t_0}{[\Delta t]} R_i \quad \text{m}^2 \cdot \text{K/W}$$

(2.3.3 - 2)

$R_{0\cdot\min\cdot N}$: minimum needful total thermal resistance, $\text{m}^2 \cdot \text{K/W}$;

t_i: indoor needful air temperature, ℃;

t_o: outdoor calculating air temperature, ℃;

R_i: inner surface thermal resistance, $\text{m}^2 \cdot \text{K/W}$;

$[\Delta t]$: allowable temperature differential between indoor air and inner surface of the envelop, ℃ or K, some examples in table 2-3.

Example: To decide the thickness of the exterior wall of a dwelling in Xian City.

[解] 设该外墙剖面如图2-18。

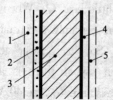

Fig. 2-18 An exterior wall section
图2-18 外墙剖面

1：内表面热阻（感热阻）$R_i = 0.115$ m²K/W；
2：石灰砂浆抹灰20，$\rho=1600$，$\lambda=0.7$，$R_2=0.0286$；
3：砖墙 $\rho=1800$，$\lambda=0.81$，厚 d 待定，$R_3=d/\lambda$；
4：水泥砂浆抹灰20，$\rho=1800$，$\lambda=0.8$，$R_4=0.025$；
5：外表面热阻（放热阻）$R_e=0.043$。

首先，算出 $R_{0 \cdot \min \cdot N}$

$$R_{0 \cdot \min \cdot N} = \frac{t_i - t_o}{[\Delta t]} R_i$$
$$= \frac{18-(-8)}{6} \times 0.115$$
$$= 0.498 \, \text{m}^2\text{K/W}$$

该外墙总热阻
$$R_0 = R_i + R_2 + R_3 + R_4 + R_e$$
$$= 0.115 + 0.0286 + d/\lambda + 0.025 + 0.043$$
$$= d/\lambda + 0.2116$$

使 $R_0 = R_{0 \cdot \min \cdot N}$，即 $d/\lambda + 0.2116 = 0.498$

已知砖墙 $\lambda = 0.81$ W/(m·K)代入得：

墙厚 $d = (0.498 - 0.2116) \times 0.81 = 0.232$ m

取 $d = 240$ mm

这就是20世纪50~80年代，西安地区居住建筑普遍采用24砖外墙的原因。其他建筑外墙也几乎都采用24砖外墙。

进入21世纪，为了节能，规范要求在20世纪80年代耗能的基准上节能50%，这就要加大热阻，减少热损失，才能做到。

(3) 节能热阻

当围护结构总热阻 $R_0 = R_{0 \cdot \min \cdot N}$ 时，

热损失 $Q = \dfrac{t_i - t_0}{R_{0 \cdot \min \cdot N}} F \cdot Z$，在此基础上节能50%，

[Solution] Presume the wall section as Fig. 2-18

Fist to find the $R_{0 \cdot \min \cdot N}$

$$R_{0 \cdot \min \cdot N} = \frac{t_i - t_o}{[\Delta t]} R_i$$
$$= \frac{18-(-8)}{6} \times 0.115$$
$$= 0.498 \ \text{m}^2 \cdot \text{K/W}$$

Here,
$$R_0 = R_i + R_2 + R_3 + R_4 + R_e$$
$$= 0.115 + 0.0286 + d/\lambda + 0.025 + 0.043$$
$$= d/\lambda + 0.2116$$

Let $R_0 = R_{0 \cdot \min \cdot N}$, i.e.

$d/\lambda + 0.2116 = 0.498$

Brick wall $\lambda = 0.81$

We can get the wall thickness d

$d = (0.498 - 0.2116) \times 0.81 = 0.232$ m

Take $d = 240$ mm

That is the reason why during 1950s to 1980s in Xi'an region the 240mm brick walls were universally to be used as the exterior walls of residential buildings (domestic buildings, dwelling buildings). Other buildings' exterior walls were also almost using 240mm brick walls.

Since coming into 21st century, the norm demands an energy saving of 50% based on the level of energy consume in 1980s, for this we must increase thermal resistance and decrease heat loss to achieve this target.

(3) Energy saving thermal resistance

When the envelop's total thermal resistance $R_0 = R_{0 \cdot \min \cdot N}$

heat loss $Q = \dfrac{t_i - t_0}{R_{0 \cdot \min \cdot N}} FZ$, based on this, to save 50% energy, two sides times by 1/2:

两边乘以 $\frac{1}{2}$：$\frac{1}{2}Q = \frac{1}{2}\frac{t_i - t_o}{R_{0\cdot min\cdot N}} F \cdot Z$

令：$2R_{0\cdot min\cdot N} = R_{0\cdot min\cdot ES}$ ES = energy saving

称 $R_{0\cdot min\cdot ES}$ 为节能所需最低总热阻

例：我们已知 20 世纪 80 年代西安住宅必需最低总热阻为 $R_{0\cdot min\cdot N} = 0.498 \doteq 0.5 m^2 K/W$，则每秒每 m^2 的热损失

$$q = \frac{t_i - t_o}{R_{0\cdot min\cdot N}} = \frac{18 - (-8)}{0.5} = 52 \text{ J}$$

若围护结构用节能热阻 $R_{0\cdot min\cdot ES} = 2R_{0\cdot min\cdot N} = 2\times 0.5 = 1.0$

那么，该热损失为 $q' = \frac{t_i - t_o}{R_{0\cdot min\cdot ES}} = \frac{18-(-8)}{1.0} = 26 \text{ J}$

达到了节能 50% 的要求。

(4) 单一砖墙保温的时代已经过去

例：按前述节能 50% 设计西安某住宅砖外墙

[解] 已知 $R_{0\cdot min\cdot ES} = 1.0 m^2 K/W$
令 $R_0 = R_i + R_w + R_e = R_{0\cdot min\cdot ES} = 1.0$
$R_i = 0.115, R_e = 0.043, R_w = d/\lambda, \lambda = 0.81 W/(m\cdot K)$

$0.115 + d/0.81 + 0.043 = 1.0$

$d = (1.0 - 0.115 - 0.043) \times 0.81 = 0.682 m$，这么厚的墙不能采用，理由：
① 太重，$1800 \times 0.682 = 1227.6 kg/m^2$；
② 结构面积太多、使用面积大大减少。

(5) 复合结构应时而生

例：将前述 682mm 砖墙改为 $2\times 120mm$ 砖墙加 60mm 绝热层组成一复合墙共 300mm 厚，既可承重盖 4～5 层楼房又可保温。那么该绝热层的热导率"λ"要多大才适

$\frac{1}{2}Q = \frac{1}{2}\frac{t_i - t_o}{R_{0\cdot min\cdot N}} FZ$

Let $2R_{0\cdot min\cdot N} = R_{0\cdot min\cdot ES}$ ES = energy saving

We term $R_{0\cdot min\cdot ES}$ the minimum needful energy saving total thermal resistance.

Example: We have known during 1980s in Xi'an, the dwellings' $R_{0\cdot min\cdot N} = 0.498 \doteq 0.5 m^2 K/W$, then the heat loss of 1s and 1m^2 is:

$$q = \frac{t_i - t_o}{R_{0\cdot min\cdot N}} = \frac{18 - (-8)}{0.5} = 52 \text{ J}$$

If the envelop with $R_{0\cdot min\cdot ES} = 2R_{0\cdot min\cdot N} = 2\times 0.5 = 1.0$

Then the heat loss $q' = \frac{t_i - t_o}{R_{0\cdot min\cdot ES}} = \frac{18-(-8)}{1.0} = 26 \text{ J}$

Just get an energy saving of 50%.

(4) The times of single brick wall as insulation have gone

Example: based on the above energy saving of 50% to design the exterior brick wall of a dwelling in Xi'an.

[**Solutian**] We have known Ro. min. ES $= 1.0 m^2 K/W$ Let $R_0 = R_i + R_w + R_e = R_{0\cdot min\cdot ES} = 1.0$, $R_i = 0.115$, $R_e = 0.043$, $R_w = d/\lambda, \lambda = 0.81 W/(m\cdot K)$,

$0.115 + d/0.81 + 0.043 = 1.0$

$d = (1.0 - 0.115 - 0.043) \times 0.81 = 0.682 m$, so much thick wall can not be used, reasons:
① Too heavy, $1800 \times 0.682 = 1227.6 kg/m^2$;
② Much structural area occupied while more usable area decreased.

(5) Composite structure arising on time

Example: To reconstruct the above 682mm brick wall into a 30cm composite wall compounded by a double 120mm brick wall filled with a 60mm insulation, which can be a

合呢？

[解] 图 2-19 为 682mm 砖墙改为 2× 120mm 砖墙中间夹 60mm 绝热层的复合墙示意图。

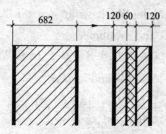

Fig. 2-19 A single brick wall reformed into a composite wall
图 2-19 单一砖墙改为复合墙

求绝热层"λ"
$$R_{0 \cdot \min \cdot ES} = 1.0$$
复合墙 $R_0 = R_i + R_b + d/\lambda + R_e$
24 墙 $R_b = 0.24/0.81 = 0.296$
绝热层热阻 $R_I = d/\lambda$，此处设定 $d = 60$mm
$R_i = 0.115, R_e = 0.043$
令 $R_0 = R_{0 \cdot \min \cdot ES} = 1.0$
i.e. $0.115 + 0.296 + 0.06/\lambda + 0.043 = 1.0$
$\lambda = 0.06/0.546 = 0.1099$W/(m·K)

沥青蛭石、沥青珍珠岩、聚乙烯泡沫板等绝热材料热导率均小于 0.1，可满足要求。

思考：绝热层放在室内一侧（暖侧）或室外一侧（冷侧），哪侧好？（以住宅为例）

不同材料组成的构件称复合结构如复合墙等。

复合构件是优势互补构件。

使用要求的多样性与材料优势效能单一性的矛盾促生了复合结构。

1) 重型复合墙

凡由混凝土或砖墙等重型构件与绝热层组合的墙体称为重型复合墙。重型复合墙可作为围护结构，承重、绝热、隔声并担当其他

bearing wall of 4~5 stories building and an insulation as well. Now, what a thermal conductivity "λ" is suitable to this insulation?

[Solution] Fig. 2-19 shows the 682mm brick wall reconstructed into a composite wall of a double 120mm brick wall filled with a 60mm insulation.

To find the "λ" of the insulation
$$R_{0 \cdot \min \cdot ES} = 1.0$$
composite wall $R_0 = R_i + R_b + d/\lambda + R_e$
24cm wall $R_b = 0.24/0.81 = 0.296$
insulation $R_I = d/\lambda$ here, set $d = 60$mm
$R_i = 0.115, R_e = 0.043$
Let $R_0 = R_{0 \cdot \min \cdot ES} = 1.0$
i.e. $0.115 + 0.296 + 0.06/\lambda + 0.043 = 1.0$
$\lambda = 0.06/0.546 = 0.1099$W/(m·K)

Asphalt vermiculite, asphalt pearlite and foamed polythene insulations, etc. their thermal conductivities are smaller than 0.1 W/(m·K), they all can satisfy the demand.

Thinking deep: the insulation attaches inside or outside of the wall, which is better? (for dwellings)

A member compounded by different materials is termed a composite member (component, element) such as a composite wall, etc.

A composite member is the member compounded by the superiorities of each other.

Use needs material with multi-function, in fact nowaday each material only has one good capability, just this contradiction has caused composite members arising.

1) Heavy composite walls

A wall compounded by a concrete or brick wall, etc. heavy elements combining with an insulation we call it a heavy composite wall.

防护任务。

Heavy composite walls can be used as envelops for load-bearing and heat insulation, sound insulation and other protective functions.

绝热层的布置

Layout of insulation

图 2-20c)绝热层在室外(冷侧),d)图中绝热层在室内(暖侧),哪个好呢?

In Fig. 2-20 c) the insulation at cold outside, in d) insulation at warm inside, which is better?

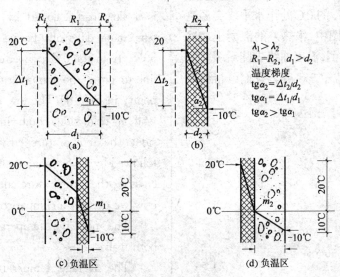

Fig. 2-20 Insulation positting in composite walls
图 2-20 复合墙中绝热层的布置

图 2-20a)中,混凝土墙暖侧表面 20℃,冷侧表面负 10℃,该墙厚为 d_1,热导率 λ_1;b)图为高绝热层,热导率 $\lambda_2 < \lambda_1$,热阻 R_2 与混凝土墙 R_1 相等,两侧表面温差与前者同,其厚 $d_2 < d_1$,则 $\mathrm{tg}\alpha_1 = \Delta t / d_1$ ℃/m 或 K/m,表明通过该墙单位距离内温度降落的度数,这就叫温度梯度。

Fig. 2-20a), A concrete wall, warm side surface temperature 20℃, cold side surface temperature negative 10℃, wall thickness d_1, thermal conductivity λ_1. In Fig. b), a high insulation, thermal conductivity $\lambda_2 < \lambda_1$, its thermal resistance R_2 equalling the concrete wall's thermal resistance R_1, its both surface temperature difference is as the same as that of the former, its thickness $d_2 < d_1$. We can see that $\mathrm{tg}\alpha_1 = \Delta t / d_1$ ℃/m or K/m, which explains that the temperature drop quantity per one unit distance is called temperature gradient.

此处,$\mathrm{tg}\alpha_1 = \Delta t / d_1$,$\mathrm{tg}\alpha_2 = \Delta t / d_2$,$\because d_1 > d_2$

Here, $\mathrm{tg}\alpha_1 = \Delta t / d_1$, $\mathrm{tg}\alpha_2 = \Delta t / d_2$ $\because d_1 > d_2$

$\therefore \mathrm{tg}\alpha_1 < \mathrm{tg}\alpha_2$,表明在混凝土墙内温度降落缓,在高绝热层内温度降落快。记住:"绝热好的材料温降快,绝热差的材料温降

$\therefore$ $\mathrm{tg}\alpha_1 < \mathrm{tg}\alpha_2$, it enunciates (explains) that in the concrete wall temperature gradient is slow, in the high insulation temperature

图 2-20 c)图中,绝热差的混凝土墙在暖侧,高绝热层在冷侧,温度曲线先缓后陡。从内外表面温差 30K 的 0℃处作水平线交温度曲线于 m_1 点;d)图中,绝热差的混凝土在冷侧,高绝热层在暖侧,温度曲线先陡后缓,0℃水平线交温度曲线于 m_2 点。我们可得到如下的比较:

墙内负温区	$d) > c)$
剖面平均温度	$d) < c)$
内部凝结危险性	$d) > c)$
维持室内温度稳定能力	$d) < c)$

结论:对连续供暖房如住宅等,绝热层布置在冷侧好。对间歇供暖房间如学校,办公楼等,绝热层布置在暖侧有利,因为外墙吸热较少,室温可较快升高。

gradient is steep. Remember: "in high insulation material temperature gradient is steep, low insulation material, slow" then we can optimize the lay out of insulation in composite wall. In Fig. 2-20c), the concrete wall (low insulation) at warm side, high insulation at cold side, the temperature curve is a slow part followed by a steep part. The temperature differential between the both sides is 30K, from the 0℃ point drawing a horizontal line to cross the temperature curve at "m_1" point; in Fig. d) The low insulation concrete wall at cold side, high insulation at warm side, the temperature curve is a steep part followed by a slow one, the 0℃ horizontal line crossing the temperature curve at "m_2" point, now we can get a comparison as follows:

The negative temperature area within both walls	$d) > c)$
The average temperature in both wall sections	$d) < c)$
Inner condensation danger	$d) > c)$
Capacity of keeping indoor temperature stability.	$d) < c)$

Conclusion: for continuous heating rooms such as dwellings, etc. the insulation at cold side is good, however, for intermittent (discontinuous) heating rooms (schools, offices, etc.) the insulation at warm side is better due to less heat absorbed by the exterior wall and indoor temperature arising quickly.

口诀比较:

"绝热差,温降缓;绝热好,温降陡"

墙内负温区	墙内正温区	剖面平均温度	内凝危险	抵抗供热波动
(d)大	(d)小	(d)低	(d)大	(d)弱
(c)小	(c)大	(c)高	(c)小	(c)强

对于间歇供热,(d)较好(室内升温快)

夹心墙

图 2-21 为几例夹心墙。两侧硬壳可用

Sandwich wall

Fig. 2-21 shows some examples of

砖、混凝土或钢筋混凝土,中填软绝热层。

sandwich walls. Their both sides are hard shells of brick, concrete or reinforced concrete and a soft insulation between them.

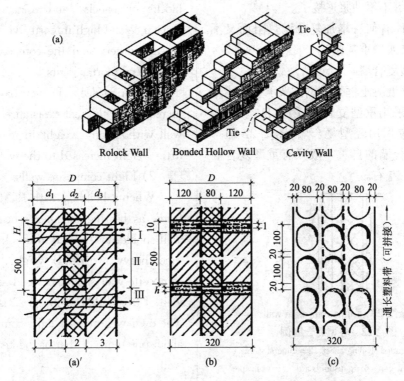

Fig. 2-21 Sandwich wall examples: (a) clay brick wall filled with air (brick lock joints or steel bar joints); (a)' brick lock joint section; (b) holed plastic tie joint section of sandwich wall; (c) holed plastic tie plane

图 2-21 夹心墙例:(a)黏土砖夹空气层(砖咬接或钢筋连接);(a)'砖咬接剖面;(b)塑料网片连接夹心墙剖面;(c)塑料网片平面

复合墙的连接

用砖咬接或不同形式的钢筋网片连接都是冷(热)桥连接,因为连接处热阻远比复合墙部位热阻小,故连接处冬季失热(夏季进热)远比复合墙部位多。

用带孔塑料带连接,连接处热阻不仅不小于甚至还大于复合墙部位,故属非冷(热)桥连接。

Joints of composite walls

Brick bridle (lock) joint and different shape steel grid joints are cold (heat) bridge joints because the thermal resistance at the joints is very smaller than that at the composite parts, through the joints winter heat loss (summer heat gain) is much more than that through the composite parts.

If using holed plastic tie, at the joints the thermal resistance is not smaller even more than that at the composite parts, so the joints are non-cold (heat) bridge joints.

无砂浆连接

当科技进一步发展,可将复合墙做成预制块材或板材,连接处用高热阻高强度粘合剂粘合,就可不用砂浆连接了。

思考:你有何法增加复合墙热阻而又不增加其重量或增重甚微?

2) 轻型复合墙

由钢筋混凝土框架或钢框架承重的建筑物,就不应采用重型复合墙,因为自重太大,此时外墙应采用轻型复合墙,图 2-22 即一例,轻型复合墙的自重一般只有重型复合墙的十分之一左右。

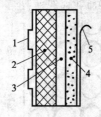

Fig. 2-22 Light composite wall
图 2-22 轻型复合墙例
1—extruded shaping sheet   2—mineral wool
3—air-tight gap   4—gypsum board   5—wallpaper
1—压型钢板   2—矿棉
3—密闭空气层   4—石膏板   5—糊墙纸

3) 高绝热单质墙

2.5 隔墙·隔空气声

思考:

为什么要隔墙?

空气、水、黄土、冰能做隔墙材料吗?

隔墙功能是分隔空间,同时应起到所处环境相应的围护作用如隔声、防水等。隔墙不承受其他构件如梁、楼板等传来的荷载。

隔墙分类

No mortar joints

As science-technology further developed, the composite walls can be precasted into blocks or panels, and using high thermal resistance and high intensity viscose to adhere to each other, and the composite walls will never need mortar joints.

Thinking deep: Do you have any way to increase the thermal resistance of composite wall without any weight increased or a very little weight increased to the wall?

2) Light composite walls

When a building with RC frame or steel frame as bearing members, heavy composite wall should not be used in this building due to much heavy own(dead) weight, now the light composite wall should be used as the exterior wall of this building, Fig. 2-22 is an example. The own weight of a light composite wall is about one tenth of that of the heavy composite wall.

3) High insulated single material walls

2.5 Partition Walls·Insulation of Airborn Sound

Thinking deep:

Why we need partitions?

Can air, water, loess and ice be the materials of partitions?

Partition duty is to divide spaces and should have the protective functions of sound insulation and waterproofing, etc. corresponding to the environment it located. Partition does not bear any other load from other elements such as beams or floors.

Types of partitions

(1) 实心砖隔断
(2) 空心砖隔断
(3) 碳化石灰板隔断
(4) 双面抹灰板条隔断

(5) 土坯砖隔断
(6) 夹空气层隔墙
(7) 夹绝热(隔声)层隔墙
(8) 充水墙
(9) 水墙
(10) 冰墙
(11) 石膏板隔墙
(12) 胶合板隔墙

图 2-23,图 2-24 为隔墙例

(1) solid brick partitions
(2) hollow tile partition
(3) carbonated lime board partition
(4) wood lath partition rendered both sides
(5) adobe partition
(6) sandwich partition with air space(gap)
(7) sandwich partition with insulation
(8) water loaded wall
(9) water wall
(10) ice wall
(11) gypsum(GYP, gyps)board partition
(12) plywood(veneer)board partition

Fig. 2-23 and Fig. 2-24 show some examples of partitions

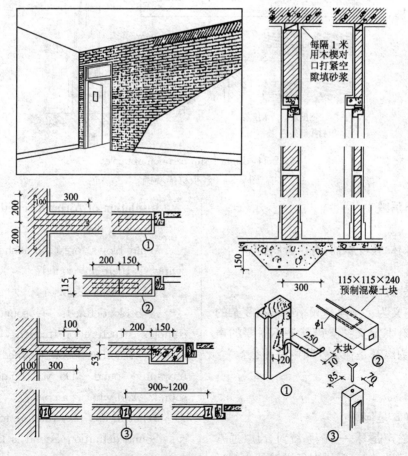

Fig. 2-23 Brick partition example
图 2-23 砖隔墙例

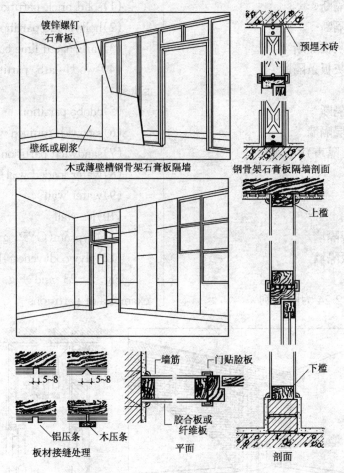

Fig. 2-24　Light board partition examples
图 2-24　轻板材隔墙例

空气声隔减

思考：

风吹森林，枝叶飘摇，有声音吗？

什么是噪声？

有书定义为："噪声是频率结构很复杂的声音"。又有书定义："噪声就是不需要的声音"。你认为声音是什么？噪声是什么？

对牛弹琴是何意？

声定义：声是耳——脑系统对介质（通常是空气）中运动的、可闻纵向行进波的主观反应。

Insulation of Airborn Sound

Thinking deep:

Wind blows forest, branches and leaves swing, is there any sound?

What means a noise?

A book defined:"A sound with very complex structure of frequency, which is a noise". Another book defined:"Noise is unwanted sound". Do you opine what is a sound? And what is a noise?

What means"play the lute to a cow"?

Sound definition: Sound is the subjective response by the ear-brain system to audible longitudinal progressive waves travelling in a

噪声就是不需要的声音

声频 f,声速 V,波长 λ 的关系:$f\lambda = V$,空气中 $V = 340$m/s,当 $f = 20$Hz,$\lambda = 340/20 = 17$m;$f = 20000$Hz,$\lambda = 340/20000 = 17$mm。

人耳听觉效应

(1)频率范围 20~20000Hz(cps)

(2)声压范围 0.0002~200μb(1000Hz)。0.0002μb 是 1000Hz 的声音刚开始引起人耳声感的声压称为 1000Hz 的"闻阈",200μb 是 1000Hz 人耳能承受的最大声压称 1000Hz 的"听觉极限"。声压闻阈与听觉极限会因频率而异,因人而异。

图 2-25 是空气载声传入人耳的示意。

声压就是声源振动器激动的介质例如空气分子产生的疏密行进波引起的对原静止压(例如大气中的大气压)P_0 的压力变动值(图 2-25)。

medium(usually air).

Noise is unwanted sound.

The relation among sound frequency(f), velocity(V) and wavelength(λ):$f\lambda = V$, in air $V = 340$m/s, when $f = 20$Hz, $\lambda = 340/20 = 17$m; $f = 20000$Hz, $\lambda = 340/20000 = 17$mm.

Auditory Response of Human Ears

(1) Frequency range 20~20000Hz(cps)

(2) Sound pressure range 0.0002 μb~200μb(F = 1000Hz). 0.0002μb is the just started sound pressure causing a sound hearing of human ears when the frequency = 1000Hz. It is termed sound pressure "threshold of hearing" of 1000Hz. 200μb is the max. sound pressure which human ears can sustain at 1000Hz, it's called "threshold of feeling" at 1000Hz. Different frequency different sound pressure threshold of hearing and feeling and different among different people.

Fig. 2-25 is showing an airborn sound transmitting into a human ear.

Sound pressure is a fluctuant pressure based on the primordially still(static) pressure P_0 (such as atmospheric pressure in air), which is caused by the dotted-dense progressive waves travelling in a medium (usually air), these waves are generated by the sound source oscillator(Fig. 2-25).

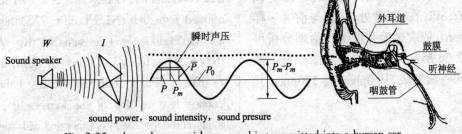

Fig.2-25 shows how an airborn sound is transmitted into a human ear.

图 2-25 空气载声传人人耳的示意

W:声功率 Acoustic power J/s; I:声强 Acoustic intensity, $I = W/m^2$;瞬时声压 Instantaneous sound pressure;P:有效声压 Effective sound pressure 又称根均方声压 P_{ms} Root-mean-square sound pressure;P_m:最大瞬时声压;Max. instantaneous sound pressure;$\bar{P}$:平均声压 Mean sound pressure;P_0:没有声波前的大气静压,Atmospherically static pressure without sound wave affecting

(3) 声压级范围 1~120 分贝

人耳听声的强弱不与声压成正比，而近似与所听声与闻阈两声压之比的常用对数值成正比。可用公式(S-1)表达如下：

(3) Sound pressure level (SPL) range 0~120 dB

A sound heared by human ears, loud or weak is not direct proportion to the sound pressure, is near proportion to common logarithm of the ratio of heared sound pressure to the pressure of threshold of hearing, it can be expressed as in formula (S-1).

$$\begin{aligned} SPL &= \log_{10} P^2/P_0^2 \quad (\text{bel}, 贝尔) \\ &= 10 \log_{10} P^2/P_0^2 \quad (dB, 分贝) \quad 1\,\text{bel} = 10\,dB \\ &= 20 \log_{10} P/P_0 \quad (dB) \end{aligned} \qquad (S\text{-}1)$$

SPL: 声压级, Sound Pressure Level, dB；
P: 听到的声压, Heared sound pressure, μb；
P_0: 闻阈声压, Sound Pressure of threshold of hearing, 0.0002μb；
在 1000Hz, At 1000Hz,

$$(SPL)_{max} = 20 \log_{10} 200\mu b/0.0002\mu b = 120\,dB$$
$$(SPL)_{min} = 20 \log_{10} 0.0002\mu b/0.0002\mu b = 0\,dB$$

0 dB 的声并不是没有声音而是刚可听到的闻阈声音。

用声压级单位 dB 进行声测量简明、方便。

(4)等响效应

不同频率不同声压级的声，当声压级在某个相应级上就会有一样的响度，这就叫等响效应。例如 $f = 1000$Hz, SPL = 10dB, 受试人听到此声(受试人均为正常听力者)用述语称此声响度级为 10phon (方)；将 f 调到 20Hz, SPL 直到 77dB 才达到与 $f = 1000$Hz, SPL = 10dB 时一样响，即 10phon；再将 f 调到 100Hz, SPL 在 30dB 处，达到与前者一样响(10phon)，依次试验并按统计物理分析可得如图 2-26 所示纯音等响曲线图。每一曲线代表同一响度级，从 0phon 到 120phon 共 121 级。

0dB sound is not no sound, it is just a sound of threshold of hearing.

Using SPL's unit dB to do sound measure is simple and convenient.

(4) Equal loudness response

Sounds with different frequencies and different sound pressure levels when their sound pressure levels at a certainly corresponding levels will have a same loudness, this is called "equal loudness response." Such as $f = 1000$Hz, SPL = 10dB, the tested people (they all have normal hearing) heard this sound and its loudness level is termed 10phon; adjusted f to 20Hz and SPL till to 77dB this sound got the loudness level as same as that of $f = 1000$Hz, SPL = 10dB, i.e. 10 phon; again, adjusted $f = 100$Hz, SPL at 30dB this sound got the same loudness level as that of the former (10 phon), going on the test one after other and with statistico-physics analysis, the testers got a group curves of pure tone loudness contours as Fig. 2-26 shows, each contour represents the same loudness level

from 0 phon to 120 phon, altogether 121 levels.

等响线图

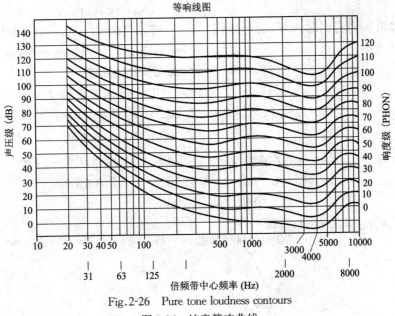

Fig.2-26　Pure tone loudness contours

图 2-26　纯音等响曲线

From Fig.2-26 we can see that the human ear sensitivity to frequency appears as:

$500 \leqslant f < 2000$Hz　sensitive　灵敏

$f = 2000 \sim 5000$Hz　more sensitive　更灵敏

$f = 3000 \sim 4000$Hz　the most sensitive, there is a "4000 dip", here human ears have a resonance. 最灵敏,对于4000Hz人耳易共振,特称"4千振谷"。

$f > 5000$Hz, $f = 300 \sim 500$Hz　obtuse　迟钝

$f < 300$Hz　most obtuse　最迟钝

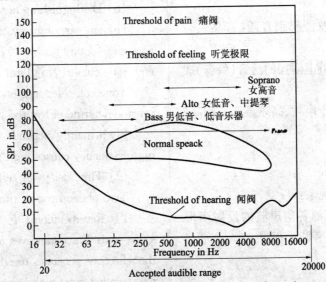

Fig.2-27　shows the generally accepted range of human audibility in terms of sound pressure level and frequency.

图 2-27　人耳通常可感受到的声压级与频率范围

(5) 声音的掩蔽与叠加

思考：

两个 100dB 的声音同时入耳，听起来该叠加声是几分贝？

简便法如下：

$L_1 > L_2$ 或 $L_1 = L_2$	ΔdB $L_1 + \Delta$dB
$L_1 - L_2 = 0, 1$dB	3dB $L_1 + 3$dB
$L_1 - L_2 = 2, 3$dB	2dB $L_1 + 2$dB
$L_1 - L_2 = 4 \sim 9$dB	1dB $L_1 + 1$dB
$(L_1 - L_2) \geqslant 10$dB	0dB $L_1 + 0$dB

(5) Sound masked and piled

Thinking deep:

Two 100dB sounds reached your ears at same time how many decibels of the piled sound you heard?

The simplified method as follows:

Multi-sound pressure levels piled

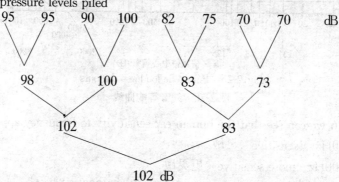

声音被掩盖在众多的噪声环境里对人的耳—脑系统实是一种保护。

Sound masked in fact is a protection for human ear-brain system in a environment of more noises.

(6) 辨别音质

人的耳—脑系统能辨别音质（音品）、音色和音调。

音色取决于基调上附加的泛音（偕音）成分。

音调主要取决于声频，高频声音音调高。

(6) Distinguishing acoustics

Human ear-brain system can distinguish acoustics (sound quality)(音质) or timbre(音品), tone colour(音色) and pitch(音调).

Tone color depends on the components of overtone(harmonic) on the keynote.

Pitch mainly depends on the frequency, high frequency causes high pitch.

(7) 延时效应

延时效应的两现象：

1) 混响：室内声源各向声波的反射声形成混响，适当的混响有利于音质。

(7) Time delay(lag) effect

Two phenomena of time delay:

1) Reverberation（混响）: indoor sound source spreads sound waves to all directions and the reflections generates reverberations, a suitable reverberation benefits the sound quality.

2 建筑科学基础

2)音程差效应:图2-28,$(a+b)>c$,直达声与反射声就有音程差效应或时差效应。

时差<50ms(毫秒)即反射音程比直达音程差在17m以内,反射声对直达声有加强作用。

时差更长,反射声很强则会形成回声。

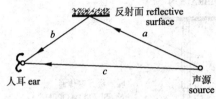

Fig.2-28　Distance travelled by sound waves
图2-28　声波音程

(8)眼耳效应

1)双耳听声可帮助辨别声源方向。

2)耳听侧面扬声器来声,而眼睛却注视着正面的讲话人,这时你脑子会认为声音来自此讲话人。

2.6　噪声控制

(1)墙体入射声能的分配

图2-29示出了空气声入射到墙体时声能的分布。

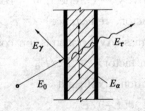

Fig.2-29　Shows the distribution of a sound energy incident on a wall
图2-29　墙体入射声能的分配

E_0:入射声能　E_0:incident sound energy;
E_γ:反射声能　E_γ:reflective sound energy;
E_τ:透射声能　E_τ:transmitted sound energy;
E_a:吸收声能　E_a:absorbed sound energy.

(2)室内外噪声入室途径

2) Different distance effect: In Fig. 2-28, if $(a + b) > c$, between direct waves and reflective waves there will cause a different distance effect(音程差效应) or an effect of equation of time(时差效应).

Equation of time<50ms(milli second) i.e. the distance travelled by the reflective waves is within 17m longer than that by the direct waves, which will strengthen the direct sound.

If the time delay is more longer and the reflective sound is strong enough, which will generate echos.

(8) Eye-ear effect

1) Two ears hearing a sound can help you to distinguish the source direction.

2) Although your ears hear the sounds coming from side loudspeaker, your eyes are looking at the frontal speaker and your mind would take the sounds coming from the speaker.

2.6　Noise Control

(1) Distribution of a sound energy incident upon(on) a wall

Fig.2-29 shows the energy distribution when an air-borne sound wave is incident on a wall.

(2) Indoor and outdoor noises transmit into rooms

图 2-30 示出了室内外噪声传入房间的状况。

Fig. 2-30 shows the indoor and outdoor noises transmitted into rooms.

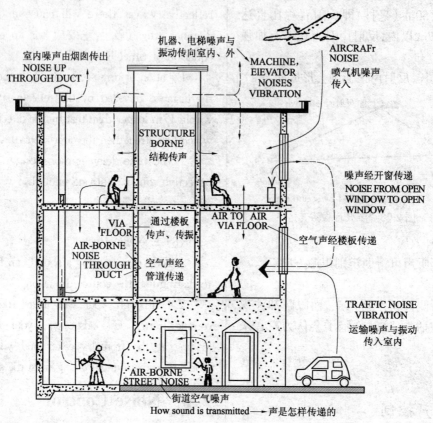

Fig. 2-30　indoor and outdoor noises transmiting ways
图 2-30　室内外噪声传递途径

有关概念

反射系数　$\gamma = E_\gamma / E_0$

透射系数　$\tau = E_\tau / E_0$

吸收(声)系数　$\alpha = (E_\alpha + E_\tau)/E_0$
$$= \frac{E_0 - E_r}{E_0}$$

工程界用"隔声量"R 或称传声损失或透射损失以分贝为单位表达构件例如墙的隔声能力。

Relative concepts

Reflectivity (reflection coefficient, reflection factor, reflectance) $\gamma = E_\gamma / E_0$

Transmissivity (transmission coefficient, transmission factor) $\tau = E_\tau / E_0$

Absorption Coefficient
$\alpha = (E_\alpha + E_\tau)/E_0$
$$= \frac{(E_0 - E_r)}{E_0} \frac{\text{未被反射的声能}}{\text{入射声能}}$$

In engineering, we use the "noise insulation factor" R or sound transmission loss (STL or TL) in dB to express the sound insulation ability of an element such as a wall.

$$R = 10\lg \frac{1}{\tau} > \text{dB}$$

$$R = 10\lg \frac{1}{\tau} > \text{dB}$$

R:隔声量,dB

τ:透射系数

例:若某墙传过入射声能的 1/1000,即 $\tau=1/1000$,则其隔声量:

$R = 10\lg \dfrac{1}{\tau} = 10\lg 1000 = 30\text{dB}$,表明该墙隔声能力 R 为 30dB。

(注:$\lg = \log_{10}$)

(3)吻合效应

图 2-31 所示,当声波以一定角度入射到隔墙时,投射的波长恰与隔墙共振弯曲波 $\lambda_w \sin\theta$ 相等时就会产生吻合效应,传入受声室的声能就会增大很多,也就是隔声能力在吻合区段会大大降低。设隔墙产生共振的横向弯曲波长为 λ_w,入射空气声波波长为 λ_i,入射角为 θ,那么,当:

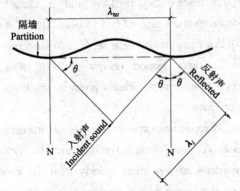

Fig.2-31　Coincidence effect
图 2-31　吻合效应

$\lambda_i = \lambda_w \sin\theta$ 时,从入射波输送的能量就会激发该墙产生如图所示共振弯曲振动。

当 $\sin\theta = 1$ 时,激发吻合效应的波长 λ_i 最长,亦即激发吻合效应的声波频率最低,称临界频率。

构件弯曲波波长是刚度的函数。刚度大的隔墙如砖墙,临界频率低(80~200Hz),不易受到激发产生吻合效应。一些轻隔断的吻合落陷可能在 2000Hz 处,正在说话频率中段,影响大,应避免,例如增加刚度,做夹心墙等。

R:noise insulation factor,dB

τ:transmission coefficient.

Example: If a wall transmitted $\dfrac{1}{1000}$ energy of the incident sound energy, i.e. $\tau=1/1000$, we can get:

$R = 10\lg \dfrac{1}{\tau} = 10\lg 1000 = 30$ dB, it is meaning that the wall sound insulation ability R is 30 dB.

(3) Coincidence effect

As Fig.2-31 showing, the coincidence effect occurs when a sound wave is incident on a partition at such an angle of incidence that the projected length of this wave is equal to $\lambda_w \sin\theta$, much more incident sound energy will be transmitted into the receiver, i.e. the sound insulation ability of the partition will be more decreased in the coincidence region. If the partition can resonate with a transverse flexural wavelength of λ_w, the wavelength of the incident airborn sound is λ_i, incident angle θ, then if:

$$\lambda_i = \lambda_w \sin\theta$$

The energy will be fed from the incident wave to motivate the partition in its resonant flexural mode as shown. When $\sin\theta = 1$, the motivating wavelength is the longest, it is meaning the motivating sound frequency being the lowest which is called critical frequency.

The wavelength of a element flexural wave is a function of stiffness (rigid). For heavy stiff partitions such as brick walls the critical frequencies are low (80 to 200 Hz), and are not easy to be motivated in resonance. Lighter partitions however may produce coincidence dips at 2000 Hz just in the middle of the speech range, which is important and should be avoided such as to increase the partitions' rigids, using sandwich walls, etc..

(4)噪声控制途径
1)治理声源
①改进工艺
例如以塑料代替钢材;锻压车间以水压代锻锤;建筑场地以挤压桩代替汽锤桩,以及拆房以胀裂代替大爆破。

②行政管理
如设立步行街;教学区禁止鸣笛;夜间禁止噪声大的施工机械运行。

2)途中消减
距离衰减与分子吸收

点声源声波在各向同性的匀质介质中均匀地各向传播,那么,声强的衰减将服从平方反比定律,距离每增一倍,声强级衰减 6dB ($10\lg\frac{(2d)^2}{d^2} = 6$ dB)。但是,大气远非匀质的各向同性介质,因此点声源声能在大气中衰减并不是理想地服从平方反比定律。

声传播途中,分子吸收是一个重要因素。声源在空气中行进,高频声比低频声衰减快得多,因为频率高空气分子振动快,吸收声能多。远处轮船笛声、喷气机声、雷声都只听到低频声就是这个缘故。

如果声音来自一线声源,例如沿公路的一列运输车,沿铁路的一列火车,平方反比定律就不适用,声音响度的衰减只与距离成反比。并取决于声源与观测者之间表面吸声特性。

(4) Noise control ways
1) Modification of sources
① Improving technology
Such as substituting steel by plastics; hammer by water press in forging workshop; hammering piles by extruding piles in building site and blasting by expanding in demolishing..

② Administration
Such as setting pedestrian mall; no bray (hoot) in school region; forbidding noisy machinery at night..

2) During propagation
Attenuation by distance and molecule absorbing

If a sound of a point source spreads out equally in all directions in a homogeneous isotropic medium, then the reduction in intensity is given by the Inverse Square Law and doubling of distance means a drop of 6 dB ($10\lg(2d)^2/d^2 = 6$ dB). But the atmosphere is far from homogeneous and isotropic. Sound energy of point source reduction in air is not ideally given by the Inverse Square Law.

Molecular absorption becomes an important factor. As sound travels through air, the higher frequencies are lost more quickly than the lower ones. Because in higher frequencies air molecular vibrating more quickly, and absorbs more sound energy. That is so we only hear the lower frequency sounds from a far steam-ship, jet aircraft and thunder.

If sound originates a line source such as a line of traffic along a road, a line of train along a railway, inverse square law does not apply, and the reduction in loudness with distance is inversely proportional to distance, and also dependent on the sound absorbing properties of the surfaces between source and observer.

公路交通噪声主要是低频谱末端声，将建筑物远离也非有效选择。建立音障，在中频及高频能很好地使声波衰减，但在低频性能较差。好在人耳对低频声反应迟钝，在一定程度上弥补了对低频声隔声难的缺点。

铁路交通噪声含空气声和地传噪声（重型卡车也有地传噪声）。空气声，如前所述可采用距离衰减与空气分子吸收以及音障（含绿化）等予以衰减。

对地传声（地层传递的振动）必须对受声建筑的基础进行处理，特别是涉及剧院设计时，更应注意。这种振动往往在闻阈以下，故常常感觉得到却听不见。

喷气机噪声像铁轨交通噪声一样，也是间歇性危害。喷气发动机产生的噪声所包含的频率成分只用简单的"A"计权网络分析是不够的。使机场远离居民区乃世界通用之法。

下列措施也常被用作陆地室外噪声的减噪措施：

乔木、灌木与花草相配合的绿化区（带）；

天然或人工的土坡；

钢筋混凝土音障墙。

3）室内隔声

室内噪声容许值举例

房　间	最宜容许值 dB(A)
广播室	20~30

Road traffic noise is mostly in the low frequency end of the spectrum, distancing the building from the roadway is usually not a viable option. Setting barrier can efficiently reduce middle and high frequency noises but can not well reduce low frequencies, just as well, the human ear is not very sensitive to those low frequencies, which to a certain degree can compensate the shortcoming of poor insulation of low frequencies.

Rail traffic noise comprises airborn sound as well as noise transmitted through the ground (heavy truck has also this noise).

Noises (vibrations) transmitted through the ground have to be dealt with in the construction of footings, and are particularly relevant in the design of theatres, very often the frequency is subsonic, and is felt rather than heard.

Aircraft noise as with rail traffic is an intermittent hazard and the sound generated by aircraft engines contains frequency components which render a simple "A" weighting analysis inadequate. Set the airport far from residential area that's a general way through the world.

The following methods are also usually used to reduce the land outdoor noises:

Area(belt) of coupling arbors and bushes to flowers and grasses;

Natural or man-made earth slope;

Reinforced concrete sound barrier.

3) Sound insulation in rooms

Examples of accepted noise values indoor

Space	Optimum dB(A)
Broadcast studios	20~30

会议室	35	Conference rooms	35
医院	35	Hospitals	35
起居室	40~45	Living rooms	40~45
卧室	35~45	Sleeping(bed) rooms	35~45
图书馆	35~45	Libraries	35~45
个人办公室	35~45	Private offices	35~45
教室	40	Schoolrooms(classrooms)	40
剧院	40	Theatres	40
大办公室	40~45	Large offices	40~45
餐馆	45	Restaurants	45
嘈杂办公室	50~60	Noisy offices	50~60

通常声源响度举例

Examples: loudnesses of normal sources

震耳欲聋声	响度	Deafening	Loudness dB
大型喷气机（头上30m）	140	Large jet aircraft (30m overhead)	140
铆接钢板(2m)	130	Riveting steel plate(2m)	130
非常吵闹声	120	Very loud	120
火车过桥(5m)		Train on bridge	
重型运输车（靠近路边石）	90	Loud heavy traffic (close to kerb)	90
闹声		Loud	
平均街道噪声	80	Average street noise	80
中等响声		Moderate	
一般办公室	60	General office	60
安静办公室	50	Quiet office	50
弱声		Faint	
卧室	30	Bedroom	30
正常呼吸（很弱声）	10	Normal breathing (very faint)	10

(4)建筑构件隔声性能

1)单一匀质隔墙

传声损失(STL)即隔声量"R"可用下式表达：

(4) Sound insulation properties of building components

1) For a single homogeneous partition Sound transmission loss (STL) (i.e. noise insulation factor "R") may express as follows:

$$R = STL = 20\lg M_0 + 20\lg f - 48 \quad dB$$

M_0:隔墙面密度,kg/m²;

f:噪声频率,H_z(cps)。

由上式可知:

a)隔墙面密度越大,隔声越好,称为质量定律;

b)噪声频率越高越易被隔减。但是想用提高墙面密度 M_0 来提高隔声量那既不合理也不经济。

举例:

12 墙	240kg/m²
24 墙	480kg/m²
1m 厚砖墙	1800kg/m²

1m厚砖墙面密度 M_0 是24墙的4倍,其隔声量只高出11.4dB!决没有人用如此厚重的隔墙作为音障。

2)轻质复合墙或夹心墙是很好的改进。

①夹空气层墙

图 2-32 夹空气层墙剖面及附加隔声值($\triangle R$ dB)曲线

M_0: Mass per unit area of the partition, kg/m²;

f: Noise frequency, Hz(hertz i.e. cps) (cps = cycles per second).

From the formula we can see:

a) the more M_0 the more STL, which is called Mass Law(质量定律);

b) the noise with higher frequency is more easy to be insulated. But if attempt to increase M_0 to increase STL that is neither reasonable nor economical. Examples:

Noise	f = 500Hz	STL = 53.5 dB
Noise	f = 500Hz	STL = 59.6 dB
Noise	f = 500Hz	STL = 71 dB

The M_0 of 100 cm thick brick wall's mass/m² is 4 times that of the 24cm one, the former's STL only 11.4dB higher than that of the latter! None uses so thick and so heavy partition to be a barrier.

2) Light(light weight) composite wall or sandwich wall is a better improvement.

①Sandwich wall with an air space(air gap)

Fig.2-23 Shows a sandwich wall with a gap and the added STL (ΔR dB)curves

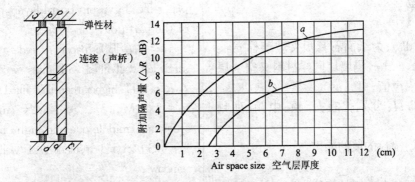

Fig.2-32 The section and $\triangle R$ dB of the sandwich wall with an air space

图 2-32 夹空气层墙剖面及附加隔声值 $\triangle R$ dB

a.
$$STL = 16\lg(M_1 + M_2)f - 30 + \triangle R \quad dB$$

STL:传声损失即隔声量,dB;

M_1, M_2:分别为两侧墙面密度,kg/m²;

f:噪声频率,Hz;

$\triangle R$:附加隔声值,dB。

b. 不要忘记当24cm厚的砖墙增厚到100cm时,其隔声量只增加了11.4dB!

c. 但是,上述夹心墙的$\triangle R$并不随空气层厚度增加而提高,其优化状态就是空气厚8~10cm时,$\triangle R = 12$dB。

思考:

你有更好的改进措施吗?

②夹吸声层墙

双层纸面石膏板或加气混凝土填以吸声材料比夹空气层墙更能提高隔声量。

例:

纸面石膏板厚12mm

M_1, M_2: double side wall's mass per unit area respectively. kg/m²;

f: noise frequency, Hz;

$\triangle R$: added STL dB.

b. Do not forget when the thickness of a 24cm brick wall increased to 100cm, its STL only increased 11.4 dB!

c. But, the above sandwich wall's $\triangle R$ is not proportional to the air gap thickness, its optimal state is air gap = 8 to 10cm, $\triangle R$ = 12dB.

Thinking deep:

Do you have any further improvement?

②Sandwich wall with absorption layer

Double paper - faced gypsum board or aeroconcrete filled with an absorption layer can get higher STL than sandwich wall with an air gap does.

Example:

Paper-faced gypsum board t 12mm

填充材料	1—1			1—11			11—11		
	L	M	H	L	M	H	L	M	H
	$\triangle R$	$\triangle R$	$\triangle R$	$\triangle R$	$\triangle R$	$\triangle R$	$\triangle R$	$\triangle R$	$\triangle R$(dB)
2.5cm 玻璃棉	6~8	6~8	6~8	9~11	9~11	9~11	12~14	12~14	12~14
5.0cm 玻璃棉	6~8	9~11	9~11	12~14	12~14	≥15	≥15	≥15	≥15
7.5cm 玻璃棉	9~11	12~14	12~14	12~14	≥15	≥15	≥15	≥15	≥15

1—1, 1—11, 11—11 表示该墙两侧纸面石膏板层数;

$\triangle R$(dB):为纸面石膏板夹玻璃棉 glass wool, glass cotton 隔墙比同样材料夹空气层墙附加的隔声值;

L, M, H:分别代表,低频、中频、高频声。

3) 有门(窗)的墙的隔声

墙和门各自传递的声能之和应等于该有门的墙组合体传递的声能这叫等传声原理。

1—1, 1—11, 11—11 are showing the paper-faced gypsum board layers on both sides of the partition;

$\triangle R$ (dB): paper - faced gypsum board sandwich wall filled with glass wool has a added $\triangle R$ of STL higher than that filled with air gap;

L, M, H: Separately marking low frequence. middle one and high one.

(3) Sound insulation of wall with door/window

The sound energy through the wall added the energy through the door together should equal to the energy through the composite

$$\tau_w S_w + \tau_d S_d = \tau_c (S_w + S_d)$$

τ_w:墙的透射系数;

τ_d:门的透射系数;

τ_c:复合墙的透射系数;

S_w:墙面积,m²;

S_d:门面积,m²。

例:(图2-33)

有一复合音障由 20m² 砖墙和一扇 2m² 门组成。设 $\tau_w = 1/10^5$,$\tau_d = 1/10^2$,计算其隔声量

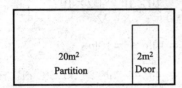

Fig.2-33 STL of a wall with a door
图 2-33 有门墙的隔声

解:首先求出 τ_c

根据等传声原理

$$S_w \tau_w + S_d \tau_d = \tau_c (S_w + S_d)$$

即

$$20 \times \frac{1}{10^5} + 2 \times \frac{1}{10^2} = \tau_c (20 + 2)$$

$$\tau_c = \frac{20 \times 10^{-5} + 2 \times 10^{-2}}{22}$$

$$= 9.2 \times 10^{-4}$$

设 R_c:复合音障隔声量;

R_w:该墙隔声量;

R_d:该门隔声量

$$R_c = 10\lg \frac{1}{\tau_c} = 10\lg \frac{1}{9.2 \times 10^{-4}} = 30.36 \text{ dB}$$

太小

改进:

(1)此处 $R_w = 10\lg \frac{1}{\tau_w} = 10\lg \frac{1}{10^5} = 50$ dB,若将 $R_w = 50$dB 提高到 $R_w = 70$dB,即 $\tau_w = \frac{1}{10^5}$ 减少到 $\tau_w = \frac{1}{10^7}$

barrier of the wall and door, which is called Equal Transmission Principle.

$$\tau_w S_w + \tau_d S_d = \tau_c (S_w + S_d)$$

τ_w: wall's transmission coefficient;

τ_d: door's transmission coefficient;

τ_c: transmission coefficient of the composite barrier;

S_w: wall area, m²;

S_d: door area, m².

Example:(Fig.2-33)

A composite barrier of a brick wall 20m² and a door 2m², if $\tau_w = 1/10^5$, $\tau_d = 1/10^2$, calculate the barrier STL.

Solution: first to find the τ_c

According to the Equal Transmission Principle

$$S_w \tau_w + S_d \tau_d = \tau_c (S_w + S_d)$$

i.e.

$$20 \times \frac{1}{10^5} + 2 \times \frac{1}{10^2} = \tau_c (20 + 2)$$

$$\tau_c = \frac{20 \times 10^{-5} + 2 \times 10^{-2}}{22}$$

$$= 9.2 \times 10^{-4}$$

Set R_c: the STL of the barrier;

R_w: the STL of the wall;

R_d: the STL of the door.

$$R_c = 10\lg \frac{1}{\tau_c} = 10\lg \frac{1}{9.2 \times 10^{-4}} = 30.36 \text{ dB}$$

This R_c is too small.

Improvements:

(1)Here, $R_w = 10\lg \frac{1}{\tau_w} = 10\lg \frac{1}{10^5} = 50$ dB, if $R_w = 50$dB increased to $R_w = 70$dB i.e. $\tau_w = \frac{1}{10^5}$ decreased to $\tau_w = \frac{1}{10^7}$

于是，$\tau_c = \dfrac{20\times10^{-7}+2\times10^{-2}}{22}$
$= 9.1\times10^{-4}$

$R_c = 10\lg\dfrac{1}{\tau_c} = 10\lg\dfrac{10^4}{9.1} = 30.4\ \text{dB} > 30.36\ \text{dB}$，仅高 0.04 dB！证明增加墙的隔声值收效甚微。

(2) 此处，$R_d = 10\lg\dfrac{1}{\tau_d} = 10\lg\dfrac{1}{10^2} = 20\ \text{dB}$

设将 $R_d = 20\ \text{dB}$ 提高到 $R_d = 40\ \text{dB}$

即 $\tau_d = \dfrac{1}{10^2}$ 减小到 $\tau_d = \dfrac{1}{10^4}$

于是 $\tau_c = \dfrac{10\times10^{-5}+10\times10^{-4}}{22} = 1.8\times10^{-5}$

$R_c = 10\lg\dfrac{1}{\tau_c} = 10\lg\dfrac{10^5}{1.8}$

$= 47.44\ \text{dB}$，比 30.36 dB 高达 17.08 dB！

结论

改善组合体隔声值的有效途径是提高薄弱构件的隔声值（传声损失）也就是降低其透射系数。

2.7 楼板·隔撞击声

2.7.1 一般讨论

思考：

(1) 为什么要楼板？什么原因促生了楼板？楼板与城市发展

(2) 下列材料可做楼板吗？

土、土坯、木材、竹子、空气、冰、石料、黏土砖、水泥、砂、卵石、砾石、钢、铁、塑料、玻璃、水、纸、混凝土、钢筋混凝土。

楼板的功能
(1)分隔垂直空间
(2)承受垂直荷载
(3)墙或框架结构的水平支撑
(4)其他相应的防护功能如隔声,防火等。

要求
(1)足够的强度含抗压强度、抗弯强度

(2)足够的刚度
刚度是抵抗变形的性能。钢丝网有好的抗拉强度,但易弯曲变形即抗弯刚度低,只能作运动员蹦弹运动用,不能作楼板。

(3)足够的防护性能
(4)经济

中、美、英如何称呼楼层?

中	美	英
一楼		底楼
二楼		一楼
三楼		二楼
……		……

Functions of Floor
(1) Partitioning vertical spaces
(2) Bearing vertical loads
(3) Horizontal supports of walls or frame structures
(4) Other corresponding protections such as protecting occupiers from noise, fire, etc.

Demands
(1) Enough strength including compressive strength(resistance) and bending strength

(2) Enough rigidity
Rigidity is the anti-deformation ability. Steel wire mesh has good tensile strength but easy bending i.e. low bending rigidity, it can be used in bouncing for sportsmen, can't be used in floor.

(3) Enough protective ability
(4) Economy

How to term the floors in China, USA and UK?

China USA	UK
the first floor	the ground floor
the second floor	the first floor
the third floor	the second floor
…	…

2.7.2 钢筋混凝土楼板

一般讨论

钢筋混凝土楼板材料是钢筋与混凝土。钢筋在钢筋混凝土楼板中主要是用作抗拉和抗剪。混凝土用水泥、砂、卵石或砾石经级配试验按优化的级配和水量拌和而成。楼板是受弯构件。在垂直荷载作用下,楼板的梁和板下部是受拉区,抗拉钢筋布置在受拉区。楼板上部是受压区,由混凝土抗压(图2-34)。混凝土抗压性能好,抗拉性能差。

2.7.2 RC floors

General

RC floor materials are steel bar and concrete. Steel bar in RC floor mainly acts as tensile reinforcement and shear reinforcement. Concrete is made of (from) cement, sand and pebble or gravel by gradation test, according to the tested gradation to set a optimal gradation, then with a set water mix them well. Floor is a member in bending. Under vertical loads the lower part of the beam and

混凝土对钢筋有握裹力,在握裹力作用下,它们共同工作,组成钢筋混凝土楼板——一种受弯构件。

slab of the floor is tensil region, upper part, compressive region. Tensil reinforcement is located in tensil region. In upper part concrete bears the pressure (Fig. 2-34). Concrete is strong in compression but weak in tension. Concrete has bond stress to bond the reinforcement (reinforcing steel, reinforcing bar, REBAR, steel bar, bar). Under the bond stress both they work together to form the RC floor—a member in bending.

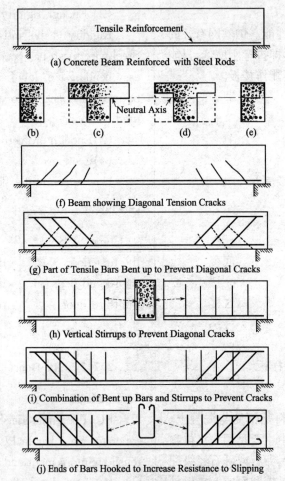

Fig.2-34　RC beam
图 2-34　RC 梁

受拉钢筋；　混凝土梁用钢筋增强；　中轴；　梁斜拉裂缝；　受拉钢筋中一部分起弯防止斜裂缝；
竖向箍筋防止斜裂缝；　钢筋端弯头成钩形增加抗滑动阻力。

Tensile Reinforcement;　Concrete Beam Reinforced with Still Rods;　Neutral Axis;
Beam showing Digonal Tension Cracks ;　Part of Tensile Bars Bent up to Prevent Pigonal Cracks;
Vertical Stirrups to Prevent Diagonal Cracks;　Ends of Bars Hooked to Increase Resistance to Slipping;

楼板还是承重墙体或框架结构的水平支撑构件,起着增加稳定性的作用(前已述及)。

楼板是节省土地构件之一,没有楼板就没有多层建筑和高层建筑。

可以说,楼板特别是钢筋混凝土楼板是促进城市发展的一种建筑构件。

2.7.3 钢筋混凝土楼板类型与施工

(1)现浇钢筋混凝土楼板

1)平板式　常用于走道、盥洗室、居住建筑及办公建筑等。

2)梁板式:

①主、次梁楼板(图 2-35),用于空间较大的框架建筑或空间较大的厅、房等处。

Floor is also the horizontal supporter of load-bearing walls and frame constructions to strengthen the stability.

Floor is one of the land saving members, without floors without multi-story buildings as well as high-rise buildings.

We may say, floor especially RC floor is a building member to help urban development.

2.7.3 RC floor types and construction

(1)Cast-in-place (cast-in(on)-site, cast-in-situ, job-placed)RC floors

1) Flat plate slabs usually are used in passageways (corridor), washrooms, residential buildings and office buildings, etc.

2)Beam-slab floors:

① Beam and girder floors are used in larger space of framing constructions or larger space halls, etc.

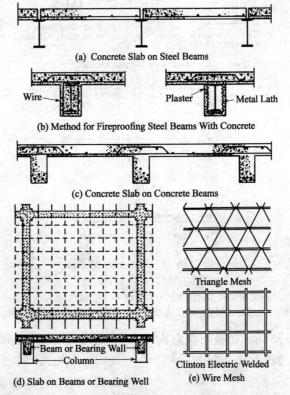

Fig.2-35　RC beam-slab floor
图 2-35　RC 梁板式楼板
(a)(钢筋)混凝土板支承在钢梁上;(b)Wire 金属线;Plaster 抹灰;
(c)(钢筋)混凝土板支在(钢筋)混凝土梁上;(d)板支在梁或承重墙上;(e)金属网

②井字楼盖(图2-36a)在这种楼板中,双向梁的截面都是一样的,常用于荷重较大的建筑如多层存车建筑、图书馆等。井字梁楼盖装饰性较好。

②Two-way rib (beam, girder) floors. In this floor the sections of the two-way rib (or beam or girder) are all the same. They are very often used in heavy-load buildings such as multi-story parking garages and multi-story libraries. The ceiling of this floor is easier to be garnished.

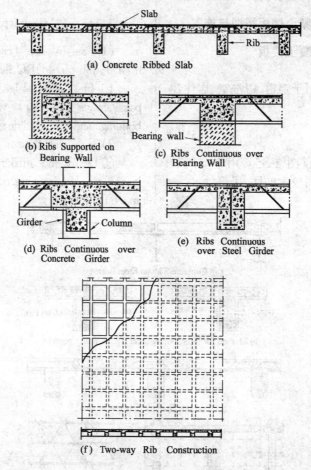

Fig.2-36(a)　Two-way rib floor
图 2-36(a)　井字梁楼盖
(a)RC密肋楼板;(b)密肋支在墙上;(c)密肋连续通过承重墙;
(d)密肋连续通过RC主梁;(e)密肋连续通过钢主梁;(f)双向密肋结构(井字梁楼盖)。

3)无梁楼盖图2-36(b),楼板布置双向抗弯钢筋。顶棚面更方便装饰,常用在百货公司等建筑物。

现浇钢筋混凝土楼板,整体性好、防水好,抗震强。地震区以及潮湿房间常采用。

3) Two-way flat-slab floor. This floor has two-way bending reinforcements. Its ceiling is more easier to be garnished. This floor is usually used in department stores, etc.

Cast - in - site RC floor has strong integrality, waterproofing quality and aseismic

现浇钢筋混凝土楼板施工程序是：

支模（钢模或塑料模）→绑扎钢筋→浇筑混凝土→振捣（用振动棒和平板振动器）→养护→脱模→必要的修补。

现浇用的混凝土有的在建房现场搅拌，但越来越多由搅拌汽车自混凝土供应厂装料→途中搅拌→建房现场，然后浇筑入模。

ability. Earthquake region and wet rooms use it frequently.

the process of cast-in-site RC floor is as follows：

Forms erected (steel form or plastic form)→ layouting and binding bars→casting concrete → tamping concrete (by poker vibrator and plate vibrator)→ curing → form removed→needful patching.

The Concrete can be mixed in building construction site, but today more and more is mixed in transit i.e. in the concrete-truck during travelling from the concrete workshop to the building construction site, then cast into the forms.

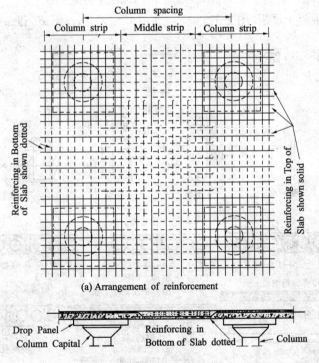

Fig.2-36(b)　Two-way flat-slab floor
图 2-36(b)　无梁楼盖
(a)钢筋布置；柱跨；柱顶楼板；中间板；板底部钢筋（虚线所示）增强；
板上部钢筋（实线所示）加强；(b)自中间板处所切剖面；柱顶板；柱冠
(a)Column spacing；Column strip；Middle strip ；Reinforcing in Bottom of Slab shown dotted ；
Reinforcing in Top of Slab shown solid ；(b)Section through middle strip；Drop panel；Column Capital

现浇钢筋混凝土的缺点是现场工期较长受气候影响较大。

克服上述缺点办法之一是采用钢板模板,在浇完混凝土后钢板留在原处不拆除,组成楼板的抗拉钢筋,处在最有利的位置并使顶棚得到平整的表面有利于表面处理。但钢材耐火不好,必须用耐火涂料予以保护。采用预制钢筋混凝土楼板也是克服现浇钢筋混凝土楼板工期长,受气候影响大的缺点的好方法。

(2)预制钢筋混凝土楼板(图 2-37)

The shortcomings of cast-in-site RC floor are longer construction period and weather affecting.

To overcome the above shortcomings, utilization of steel plat forms is one of the ways. After concrete casted on the forms which never be removed, left in the place to become the tensile reinforcement at the most optimal position, and provide a uniform flat surface over the whole ceiling benefiting the surface treatment. But steel fire resistance hour is very low, a fire resisting paint is needful. Using precast RC floor is also a good way to overcome the cast-in-site RC floor shortcomings of long construction period and weather affecting.

(2) Precast RC floors(Fig. 2-37)

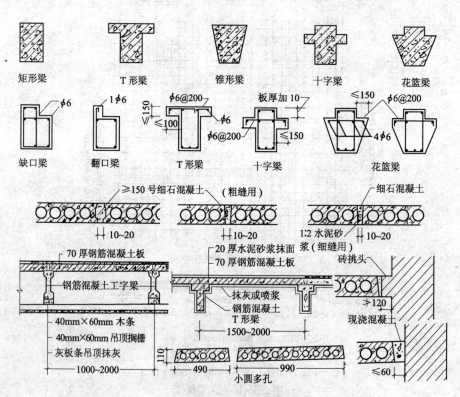

Fig.2-37 Precast RC floors(单位:mm)
图 2-37 预制钢筋混凝土楼板

图 2-37 为多种预制钢筋混凝土楼板及梁截面例图。

梁按截面形状有：矩形梁、T 形梁、锥形梁、十字梁、花篮梁等。后四种梁截面都是为了使截面不要过高而受压区抗压混凝土面积又不减少而设计的。（现浇梁也可采用）。板按截面分有：实心板、单孔板、多孔板等，板宽自 390mm 到 1200mm 甚至住宅一间房间那么大，板的跨度自 1800mm 到 6000mm，板厚：实心板 60～80mm，空心板 110～250mm。

2.7.4 预应力钢筋混凝土楼板

我们已经知道钢筋混凝土楼板是受弯构件，其下部是受拉区，如果在受拉区预先加以恰当的压应力，那么在垂直荷载作用下这种预应力板比普通钢筋混凝土楼板（非预应力楼板）会显示下列优势：产生的挠度（向下弯曲的程度）要小；出现弯曲裂缝很少，也就是抵抗变形的能力（刚度）提高了；比普通楼板厚度可减少，例如普通多孔楼板厚度 140mm，预应力多孔楼板只要 110mm 就够了，做成双向预应力实心板（可做到住宅一间房间那么大）只要 80mm 厚就够了，那么对多层建筑来说，房屋的总高度就可降低，具有省材、省工、省时，省钱的优点。因此，施工水平较高的地区例如城市采用预制楼板时都普遍采用预应力钢筋混凝土楼板。

Fig. 2 - 37 shows multiform (various) section examples of precast RC floors and beams.

According to the section shapes, beams have rectangular beams, T-shaped, cone, cross-shaped and flower - basket shaped, etc. The latter four sections are designed for reducing the beam height while no reducing the compression concrete area in compressive region(They can be used in cast-in-site RC floor too).

According to the section shapes, slabs have solid slabs, single hollow slabs, multi-hollow slabs, etc. the slab widths from 390mm to 1200mm even to a width of a room of dwellings, slab span from 1800mm to 6000mm, slab thickness: solid slab, 60 ～ 80mm, hollow slab, 110～250mm.

2.7.4 Prestressed RC floors

We have known RC floors are members in bending, the lower part of the member is tensile region. If the tensile region is given a suitable precompressive force, under the vertical loads this prestressed slab will appear the following advantages than the ordinary (non - prestressed) one does: less deflection (downward amount of deflection); few bending cracks that means the capacity of anti - deformation (i.e. rigid) increased; the thickness of prestressed slab is thinner than that of the ordinary one such as if an ordinary multi - hollow slab thickness is 140mm, the thickness of the prestressed one only 110mm is enough, the thickness of a two-way prestressed solid slad (its dimension can be made as large as a room of a dwelling) only 80mm is enough, so, for multi-story building the total height of

楼板受拉区的预压力怎样预先加上去呢？举一例如下：

设有一混凝土地面预制场地 100m 长，50m 宽，先涂上隔离剂（脱模剂）（例如皂液即肥皂水）以免楼板与混凝土地面粘结→按板宽支边模→排好高强钢筋，$\phi 5 \sim \phi 6$，长 100m + 张拉锚固用长度两端各 1m 左右，钢筋间距 75～85（按设计）→用张拉机对钢筋逐根张拉到设定拉力→两端锚固→用 30mm × 30mm × 10mm（厚）水泥砂浆块将钢筋垫离隔离层 10mm（即保护层）→浇筑混凝土→平板振动器振实→抹平表面→养护至 75%以上设计强度→松开锚固、钢筋回缩（钢筋是弹性材料，在弹性范围内，张拉力释放后将恢复原状），通过钢筋与混凝土之间的握裹力作用，混凝土就受到了预压应力。

预制—现浇相结合钢筋混凝土楼板（装配整体式楼板）

采用预制钢筋混凝土梁、板、柱，在建房现场浇筑混凝土将它们连接在一起，这种楼板既省模板，省一定的工期又有相当的整体性。在某些多层公共建筑有时会被采用。

预制钢筋混凝土楼板接缝见图 2-37

the building can be reduced which causes the benefits of the saves of material, labour, time and money. For this, in the region of heigher construction level, such as in cities, when using precast floors the prestressed RC slabs are univesal to be used.

How to precompress the tensile zone of the slab? An example: Let a precast site with a concrete ground of 100m (length) × 50m (width), first, painting a layer of release agent (such as soapsuds) on the concrete ground → side forms erected → layouting well bars, $\phi 5 \sim \phi 6$, the whole length = 100m + 2m of anchoring (each end 1m), @ (pitch between the bars) 75 ~ 85 (according to design) → stretching the bars one after another to a set tension → both ends stretched → underlaying the bars to make a protection layer with cement-sand mortar blocks (each dimension 30mm × 30mm × 10mm (thickness)) → casting concrete → tamped concrete by plate vibrator → curing the concrete to a intensity of ≥75% of the design intensity → stretches released → bars shrunken (bar is a elastic material, within elastic range, after stretching force released, the bar will shrink to reset condition) and the concrete being precompressed through the bond stress action between concrete and bar.

RC floors of precast combined with cast-in-site

Precast RC beams, slabs and columns joined together with concrete in building place which can save forms, a certain construction period and provide a suitable integratity. These floors some times can be used in some multi-story public buildings.

The joints of precast RC floors see fig. 2-37

2.7.5 撞击声隔减

钢筋混凝土楼板很重,有足够的隔空气声能力,但隔撞击声能力弱。

下列方法常用于提高其隔撞击声能力:

(1)软鞋;(2)地毯,必须经常保持清洁并消毒,否则,细菌丛生;(3)弹性垫;(4)隔声吊顶。

2.7.6 阳台、雨棚及其绿化效益

阳台使用得好可提供:1)主房间不宜的晾晒空间;2)开阔的视景并便于使用者与室外环境的信息交流;3)阳台绿化,增强生态正效应(在光合作用下吸收 CO_2,提供 O_2 以及美色香味);4)节能空间(以冬季为例,将阳台封闭,利用温室效应可节能 30%～40%)。

雨棚是设在出入口遮挡雨雪便于人们出入的构件。图 2-38 为杭州机场长廊式雨棚。雨棚上轻质种植爬山虎等植物有利于夏天隔热、防回溅水、生态正效应。

2.7.5 Insulation of impact sound

RC floor has enough insulation of airborn sound due to its heavy mass but weak insulation of impact sound.

The following methods are usually used to increase insulation of impact sound in RC floors:

(1) soft shoes; (2) carpet, it must be usually cleaned and sterilized. If not, much germs will grow; (3) spring underlay; (4) sound insulation suspended ceiling.

2.7.6 Balcony, canopy(weather shed) and their greening benefits

If you use the balcony well that can provide: 1) an airing space which is not suitable in main room; 2) a spacious sight and the information exchange between the users and the outdoor environments; 3) balcony greening to increase positive eco-effect (under photosynthesis, the plants can absorb CO_2, supply O_2, beautiful color, sweet smell.); 4) energy saving space due to green house effect. In winter, close the balcony can save energy 30%～40%.

Canopy is a building element set at the entrance of a building to shade raining or snowing so as to let people conveniently go in/out. Fig.2-38, the gallery canopy of Hangzhou airport.

Planting Boston ivy, etc. by a light material planting layer on conopy benefits: winter & summer insulation; cancelling rebounding water; positive eco-effect.

Fig. 2-38　The gallery canopy of Hangzhou airport
图 2-38　杭州机场长廊式雨棚

2.8　楼梯

楼梯是建筑中上下交通设施之一，其他上下交通设施见(1)~(7)列名：

(1)电梯(英)，电梯(美)；
(2)自动扶梯；
(3)坡道；
(4)台阶(图 2-44)；
(5)爬梯(图 2-39)；
(6)爬杆；
(7)爬绳。

我们将重点讨论钢筋混凝土楼梯。

钢筋混凝土楼梯

功能：平时上下交通；紧急疏散；锻炼身心。

2.8　Stairs

Stair is one of the up and down traffic tools, other up and down tools are listed as (1)~(7):

(1)Lift(England)elevator(America);
(2)Escalator(moving staircase, moving stairway, travelling staircase, automatic staircase);
(3)Ramp(slope ramp, slopeway);
(4)Steps(steps leading upto a house);
(5)Climbing stairway;
(6)Climbing pole;
(7)Climbing rope.

We'll focus on RC stairs

RC stairs

Functions: Daily up and down traffic; urgent escape; making body and heart strong.

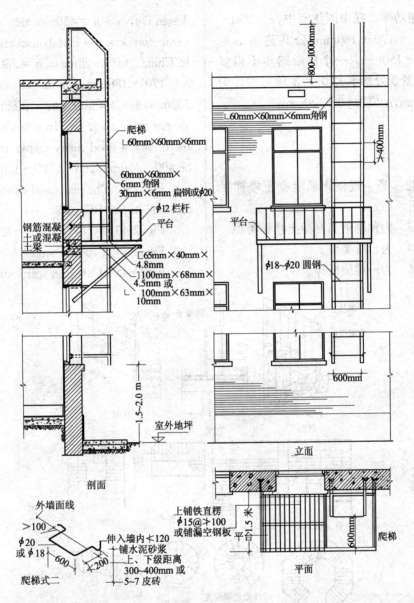

Fig.2-39 Climbing stairway

图 2-39 爬梯

图 2-40 楼梯的型式
图 2-41 楼梯的组成

梯段,休息平台(Fig.2-40);栏杆;扶手;踏板;踢板;相邻踢板面净距;踏板凸缘;梯段斜梁;明步楼梯(明楼梯);暗步楼梯(暗楼梯);端柱;踢脚板或底脚线。

设"a" = 踏板宽,"h" = 踢板高,设计中有一规则即:$a + h = 450mm$,这符合一般人

Fig.2-40 shows various forms of staircases

Fig.2-41 Component of a stair

Flight, Landing (Fig 2-40); Baluster; Rail; Tread; Riser; Run; Nosing; String; Open String Stairs (Open Stairs); Closed String Stairs (Closed Stairs); Newel Post; Skirting or Base。

Set "a" = the width of a tread, "h" = the rise i.e. the height of a riser, there is a rule in

的新陈代谢功率。在中国住宅中，$a = 280 \sim 270$mm，$h = 170 \sim 180$mm；公共建筑 $a = 300$mm，$h = 150$mm。一个梯段踏步不得多于18级。景观处踏步太多时，a值可设计为 $400 \sim 450$mm，h值为 $80 \sim 120$mm。

图 2-42 是一现浇钢筋混凝土楼梯平面及剖示图

图 2-43 为预制钢筋混凝土楼梯类型

图 2-44 为台阶类型。

图 2-45 为一螺旋楼梯。

design i.e. $a + h = 450$mm that's suitable to a common person's metabolism energy power. In China, for dwellings, $a = 280 \sim 270$mm, $h = 170 \sim 180$mm; for public buildings, $a = 300$mm, $h = 150$mm. On a flight, the risers are not more than 18. In a landscape site, if there have a good many steps, the "a" may be $400 \sim 450$mm and "h" $80 \sim 120$mm。

Fig. 2-42 The plans and cutaway drawing of a cast-in-site RC stair.

Fig. 2-43, Types of precast RC stairs.

Fig. 2-44, Types of steps

Fig. 2-45 Cast-iron spiral stair

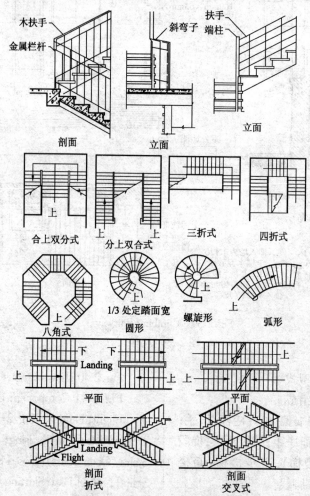

Fig.2-40 Various forms of staircases

图 2-40 楼梯的型式

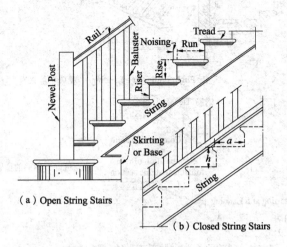

Fig. 2-41 Component of a stair
图 2-41 楼梯的组成

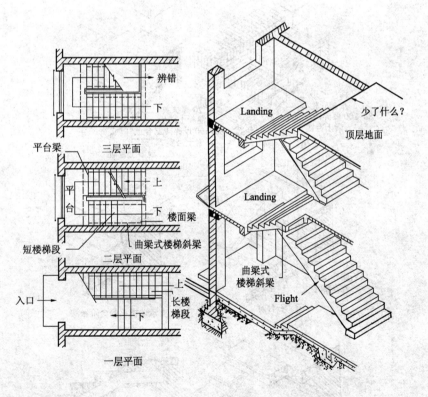

Fig. 2-42 The plans and cutaway drawing of a cast-in-site RC staircase
图 2-42 现浇 RC 楼梯间平面及剖示图

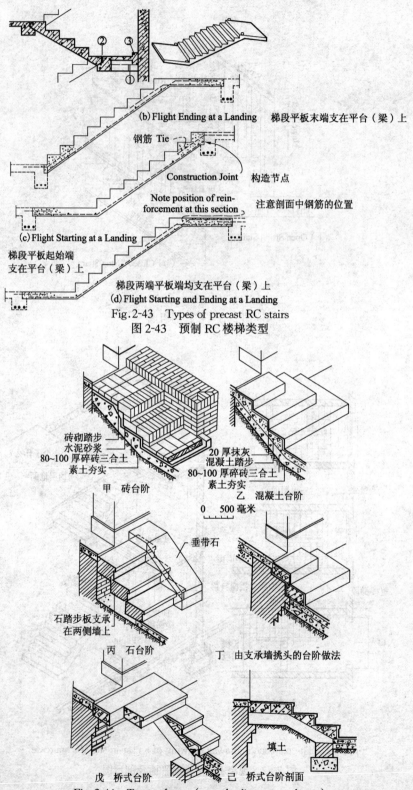

Fig.2-43 Types of precast RC stairs
图 2-43 预制 RC 楼梯类型

Fig.2-44 Types of steps(steps leading upto a house)
图 2-44 台阶类型

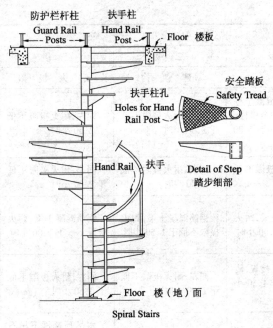

Fig. 2-45 Cast-iron spiral stairs
图 2-45 铸铁螺旋楼梯

2.9 建筑防火

水火无情

例:公元 2000 年,中国发生了 18 万多次火灾,大量房屋被烧毁,经济损失达到 258 亿元人民币。

建筑物耐火等级、建筑构件的燃烧性能与耐火时限,见前面表 1-2(Table1-2)。

表 2-4 列出了建筑物耐火等级和构造举例。

2.9 Building Fireproofing

Floods and fires have no any mercy to anything.

Example: In China in 2000, more than 0.18 million times conflagrations happened, a lot of houses were burned down and 25.8 billion Yuan RMB lost in economy.

Buildings' FRR, the combustibility and fire resistance hour of building members see Table 1-2.

Table 2-4 shows the fire resistance rating of buildings and construction examples.

表 2-4 建筑物耐火等级及构造举例

构件名称	耐 火 等 级			
	一级	二级	三级	四级
	建 筑 构 造 及 耐 火 极 限			
承重墙与楼梯间墙	砖石材料、混凝土、毛石混凝土、加气混凝土、钢筋混凝土。耐火极限不低于 3.00 小时	同左,耐火极限不低于 2.50 小时	同左,耐火极限不低于 2.50 小时	木骨架两面钉板条抹灰、苇箔抹灰、钢丝网抹灰、石棉水泥板,耐火极限不低于 0.50 小时

续表

构件名称	耐火等级			
	一级	二级	三级	四级
	建筑构造及耐火极限			
支承多层的柱	砖柱、钢筋混凝土柱或有保护层的金属柱,耐火极限不低于3.00小时	同左,耐火极限不低于2.50小时	同左,耐火极限不低于2.50小时	有保护层的木柱,耐火极限不低于0.50小时
支承单层的柱	同上,耐火极限不低于2.50小时	同上,耐火极限不低于2.00小时	同上,耐火极限不低于2.00小时	无保护层的木柱
梁	钢筋混凝土梁,耐火极限不低于2.00小时	钢筋混凝土梁,耐火极限不低于1.50小时	钢筋混凝土梁,耐火极限不低于1.00小时	有保护层的木柱,耐火极限不低于0.50小时
楼板	钢筋混凝土楼板,耐火极限不低于1.50小时	同左,耐火极限不低于1.00小时	同左,耐火极限不低于0.50小时	木楼板下有难燃烧体的保护层,耐火极限不低于0.25小时
吊顶	钢吊顶搁栅下,吊石棉水泥板、石膏板、石棉板或钢丝网抹灰,耐火极限不低于0.25小时	木吊顶搁栅下吊钢丝网抹灰、板条抹灰,耐火极限不低于0.25小时	木吊顶搁栅下吊石棉水泥板、石膏板、石棉板、钢丝网抹灰、板条抹灰、苇箔抹灰、水泥刨花板,耐火极限不低于0.15小时	木吊顶搁栅下吊板条、苇箔、纸板、纤维板、胶合板等可燃物
屋顶承重构件	钢筋混凝土结构,耐火极限不低于1.50小时	钢筋混凝土结构,耐火极限不低于0.50小时	无保护层的木梁	无保护层的木梁
楼梯	钢筋混凝土楼梯,耐火极限不低于1.50小时	钢筋混凝土楼梯,耐火极限不低于1.00小时	钢筋混凝土楼梯,耐火极限不低于1.00小时	木楼梯
框架填充墙	砖、轻质混凝土砌块、硅酸盐砌块、石块、加气混凝土构件、钢筋混凝土,耐火极限不低于1.00小时	砖、轻质混凝土砌块、硅酸盐砌块、石块、加气混凝土构件、钢筋混凝土板,耐火极限不低于0.50小时	砖、轻质混凝土砌块、硅酸盐砌块、石块、加气混凝土构件、钢筋混凝土板,耐火极限不低于0.50小时	木骨架两面钉石棉水泥板、石膏板、水泥刨花板、钢丝网抹灰、苇箔抹灰、板条抹灰,耐火极限不低于0.25小时
隔墙	砖、轻质混凝土砌块、硅酸盐砌块、石块、加气混凝土构件、钢筋混凝土板,耐火极限不低于1.00小时	砖、轻质混凝土砌块、硅酸盐砌块、石块、加气混凝土构件、钢筋混凝土板,耐火极限不低于0.50小时	木骨架两面钉石膏板、石棉水泥板、钢丝网抹灰、板条抹灰、苇箔抹灰,耐火极限不低于0.50小时	木骨架两面钉石棉水泥板、石膏板、水泥刨花板、钢丝网抹灰、苇箔抹灰、板条抹灰,耐火极限不低于0.25小时
防火墙	砖石材料、混凝土、加气混凝土、钢筋混凝土,耐火极限不低于4.00小时	砖石材料、混凝土、加气混凝土、钢筋混凝土,耐火极限不低于4.00小时	砖石材料、混凝土、加气混凝土、钢筋混凝土,耐火极限不低于4.00小时	砖石材料、混凝土、加气混凝土、钢筋混凝土,耐火极限不低于4.00小时

Table 2-4 Buliding Fire Resistance Rating and Construction Examples

Member Name	Fire Resistance Rating(FRR)			
	1st	2nd	3rd	4th
	Building Construction and Fire Resistance Hour(FRH)			
Bearing walls and staircase walls	Brick, stone, concrete, rubble concrete, aerocrete, RC. FRH≥3.00h	As the same as the left. FRH≥2.50h	As the same as the left.	Wood frame both sides lathing mortared, reed matting mortared, metal lathing mortared, asbestos cement board. FRH≥0.50 h
Columns of multi-story buildings	Brick columns, RC columns and metal columns protected. FRH≥3.00h	As the same as the left. FRH≥2.50	As the same as the left.	Wood posts protected. FRH=0.5h
Columns of single story buildings	Ditto. FRH≥2.50h	Ditto. FRH≥2.00h	Ditto. FRH≥2.00h	No protected wood posts
Beams	RC beams. FRH≥2.00h	RC beams. FRH≥1.50h	RC beams. FRH≥1.00h	Wood beams protected. FRH≥0.50h
Floors	RC floors. FRH≥1.50h	As the same as the left. FRH≥1.00h	As the same as the left. FRH≥0.50h	With nonflammables as protection beneath the wood floor. FRH≥0.25h
Ceilings	Asbestos cement board, gypboard, asbestos board or metal lathing mortared hung beneath steel ceiling joists. FRH≥0.25h	Metal lathing mortared, lath mortared hung beneath wood ceiling joists. FRH≥0.25h	Asbestos cement board, GYP board, asbestos board, metal lathing mortared, lath mortared, reed matting mortared or cement shaving board hung beneath wood ceiling joists. FRH≥1.50h	Flammables: laths, reed matting. paperboard, fibreboard or plywood hung beneath wood ceiling joists.
Roof bearing members	RC structures. FRH≥1.50h	RC structures. FRH≥0.50h	No protected wood beams.	As the same as the left.
Stairs	RC stairs. FRH≥1.50h	RC stairs. FRH≥1.00h	As the same as the left.	Wood stairs
Walls filled in frames	Brick, light concrete blocks, silicate blocks, stones, aerocrete, RC. FRH≥1.00h	As the same as the left. FRH≥0.50h	As the same as the left	Wood frame both sides asbestos cement board, GYP board, cement shaving board, metal lathing mortared, reed matting mortared, lath mortared. FRH≥0.25h
Partition walls	Ditto.	Ditto.	Wood frame both sides GYP board, asbestos cement board, metal lathing rendered, lath rendered, reed matting rendered. FRH≥0.50h	Ditto FRH≥0.25h
Fire walls	Brick, stone, concrete, aerocrete, RC FRH≥4.00h	As the same as the left.	As the same as the left.	As the same as the left

(1) 何谓建筑构件耐火时限？

按照国际标准化组织(ISO)的标准对构件进行测试。该构件耐火时限意义如下：

1) 从构件单面受一标准火作用开始直至其失去支承能力时止所经历的小时数；

2) 或直至出现穿透性裂缝时止所经历的小时数；

3) 或直至试件背面温升达 220℃ 所经历的小时数。

(2) 提高耐火时限的主要方法

1) 增厚

例：粘土砖墙

    12cm, FRH = 2.5h

    24cm, FRH = 5.0h

混凝土保护层 10mm, FRH = 1.0h

            20mm, FRH = 2.0h

2) 用绝热体隔离，如炼钢厂钢吊车梁下悬挂石棉水泥板隔离高温烟火就是一例。

3) 防火涂层

一种用绝热材料如膨胀蛭石或膨胀珍珠岩与一种无机粘结剂拌成的防火涂料，涂在钢构件上，厚18mm，耐火时限可达1.5h；厚25mm，耐火时限可达2.0h。（钢构件无防火涂层，FRH 仅 0.25h）。

(3) 防火分区

防火分区的目的就是用水平的或垂直的防火构件将建筑空间分隔开，以便阻止或延

(1) What means fire resistance hour (FRH) of building members?

According to the standards of the ISO (International standardization Organization) to test the building member, the meaning of FRH of the member is that:

1) The undergone hours from the beginning of a standard fire to firing one side of the member till its bearing capacity lost;

2) or till a penetrated crack appeared;

3) or till the temperature of the member reverse upto 220℃.

(2) Main methods of Increasing FRH

1) Width increased

Examples: clay brick walls

    12cm, FRH = 2.5h

    24cm, FRH = 5.0h

Concrete protective coating (finish, course or layer) 10mm, FRH = 1.0h; 20mm, FRH = 2.0h.

2) Isolated with isolator (isolater).

An example, in steel mill (steel works) under the steel crane beams (crane girders) usually hanged asbestos-cement boards isolate the high temperature smoke and fire.

3) Fire resistance paint (painting, coating)

There is a fire resistance paint made by an insulation material such as expanded vermiculite or expanded perlite mixed with an inorganic bond, paint it on a steel member, thickness 18mm, FRH = 1.5h; 25mm, FRH = 2.0h, (The steel member without the fire resistance paint, its FRH only 0.25h).

(3) Fire Compartment

The aim of fire compartment is to divide the building spaces with vertical or horizontal

缓火势蔓延并便于人们紧急疏散。

水平防火分区
分区面积
允许最大面积
耐火等级的1~2级的普通建筑 2500m²
  三级耐火   1200m²
  四级耐火   600m²
高层建筑
Ⅰ类高建   1000m²
Ⅱ类高建   1500m²
高层建筑地下室   500m²

(4)分隔构件
1)防火墙,常用非燃烧材料如黏土砖或混凝土块材建成。

例如240mm的黏土砖墙,耐火时限5小时,已足够满足防火要求。

防火墙应有自己的基础。当屋顶为木屋架时,防火墙应高出该屋面500mm。框架建筑中,防火墙应砌在梁上,并由柱中伸出 $\phi6$ 钢筋至砖墙缝中加以连接,伸入长度不小于500mm,入缝钢筋上下垂直距离不大于500mm,柱每侧入缝钢筋不应小于两根。

2)防火门
图2-46所示为多种防火门剖面及其耐火时限以及多种隔声门及其STL(dB)。

图2-47所示为单页及双页平开防火门细部构造。

fire members so as to stop or delay the fire spreading and let people urgently escape.

Horizontal Fire Compartment
Compartment area
Maxmum allowed area
Common buildings of first or second FRR
            2500m²
  of third FRR   1200m²
  of forth FRR   600m²
High-rise buildings
 Ⅰ type high-rise buildings 1000m²
 Ⅱ type high-rise buildings 1500m²
 Basements of high-rise buildings 500m²

（4）Isolaters(Isolators)
1)Fire walls
Fire walls are usually made by noncombustible (incombustible) materials such as clay bricks or concrete blocks.

For an example, a clay brick firewall of 240mm thickness, FRH = 5h that's very enough to satisfy the fire resistance demands.

Firewall should has its own foundation. When the roof with wood trusses the firewall should overtop the roofing 500mm. In a framed building, firewall should be laid on the beams and jointed by the columns with $\phi6$ bars from column extended into the brick joints not shorter than 500mm, the vertical pitch between the bars not larger than 500mm. The bars extended into the brick joints at the same level should not be fewer than two pieces from each side of the columns.

2)Fire doors
Fig.2-46 shows various sections of fire doors and their FRH and various sound insulation doors with their STL(dB)

Fig.2-47 shows the swinging and double swinging fire door details.

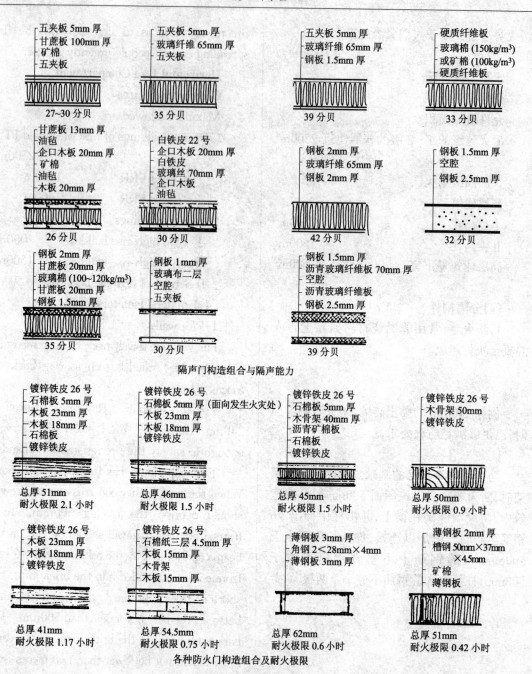

Fig. 2-46 Various sections of fire doors and their FRH and various sound insulation doors with their STL(dB)

图 2-46 多种防火门剖面构造及耐火时限和多种隔声门及其隔声量(dB)

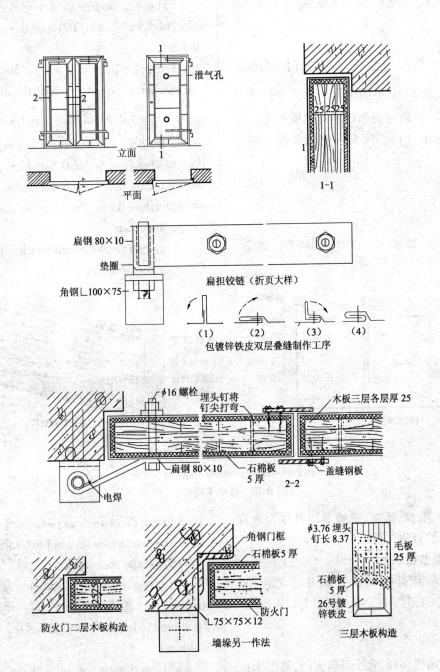

Fig. 2-47 The construction of a swinging fire door
图 2-47 平开防火门构造

防火门净宽可按每100人宽1.0m确定。

例：一教室，学生50，教师1，门净宽应是 $5 \times \frac{1}{100} = 0.51m$，不能采用。按日常使用及搬物所需，该教室每扇门净宽的取0.9m，安装了两扇平开门。防火门应向疏散方向开启。

图2-48所示为一自动关闭防水门构造

The net width of a fire door can be calculated with each 100 persons 1.0m net width.

Example: In a classroom, 50 students, one teacher. The net width of the door should be $51 \times \frac{1}{100} = 0.51m$, this is not available, as usual, for daily use and carry, each classroom door needs net width of 0.9m. Here we fix on two swinging doors, each net width of 0.9m. swinging fire doors should open towards the escape direction.

Fig. 2-48 shows an automatically shutting fire door construction.

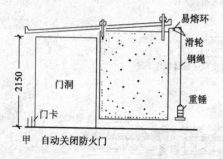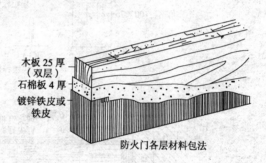

Fig. 2-48 Automatically shutting fire door

图2-48 自动关闭防火门

楼板、阳台、雨棚、挑檐等均可作为垂直防火分区的分隔构件

(5)疏散楼梯间

1) 开敞楼梯间，不设门。适用：①建筑≤11层单元式住宅；②高＜24m的工业与民用建筑。

2) 封闭楼梯间，有平开防火门，适用于：

①医院、疗养院的病房楼；②有空调系统的多层旅馆；③≥6层的公共建筑；④7～8层塔式住宅；⑤12～18层单元住宅；⑥＜11层通廊住宅；⑦高≤32m的Ⅱ类高层建筑。

Floors, balconies, canopies, overhangs, etc. can be the isolators for vertical fire compartments.

(5) Escape staircases

1) Open staircase, without any door, suitable to: ① buildings ≤ 11 stories unit dwellings; ② height＜24m industrial and civil buildings.

2) Closed staircase must have swinging fire doors, this staircase is suitable to: ① the ward building in hospital or sanatorium; ② multi-story hotels with air conditioning system; ③ ≥ 6-story public buildings; ④ 7～8-story tower dwellings; ⑤ 12～18-story unit dwellings; ⑥ ＜11-story full-length

3) 防烟楼梯间

图 2-49 为一带开敞通风前室的防烟楼梯间例。当该楼梯间带有封闭前室时则需要设通风道将室内烟气排走。

passage dwellings; ⑦ height ≤ 32m Ⅱ type high-rise buildings.

3) Smoke prevention staircases (smokeproof staircases)

Fig.2-49 is an example of smokeproof staircase with an open air vestibule. When the staircase with a closed vestibule, a tunnel is necessary to exhaust the indoor smoke.

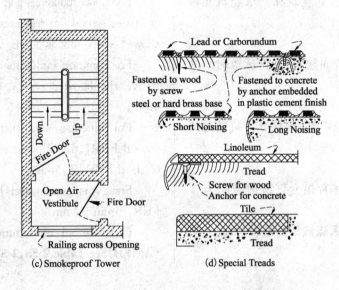

Fig.2-49 Smokeproof staircase with an open air vestibule

图 2-49 带开敞通风前室的防烟楼梯间

同一层楼疏散楼梯梯段净宽按 100 人 1.0m 计算。最多人数按本层或本层以上任一层最多的人数计算。

The net width of the flights at the same floor can be calculated with 100 persons by 1.0m. The most person number according to this floor or one of the upper floors should be taken in calculation.

(6) 安全疏散时间

定义:

普通建筑:自疏散开始至到达室外所经历的时间(以分计);

高层建筑:自疏散开始至到达最近的封闭楼梯间或防烟楼梯或避难所经历的时间(以分计)。

(6) Safe escape time

Definition:

In common buildings: From escape start till reached outdoor. the undergone minutes we call escape time;

In high-rise buildings: From escape start till reached the nearest closed staircase or smokeproof staircase or refuge, the undergone minutes we call escape time.

根据调查研究,房屋着火后 5~8 分钟将达到猛烈燃烧阶段,产生大量 CO,CO_2 和烟气并严重缺 O_2,故安全疏散时间即允许的疏散时间不应超过 5~8 分钟。

1~2 级耐火公共建筑和高层建筑允许疏散时间为 5′~7′;

3~4 级耐火普通建筑,允许疏散时间为 2′~4′;

1~2 级耐火电影院、剧院允许疏散时间为 2′;3 级为 1.5′;

1~2 级耐火体育馆允许疏散时间为 3′~4′。

(7)允许的最远疏散距离见表 2-5a,2-5b

Based on investigations and researches, when a building is on fire, the flash over fire will occur after 5~8 minutes, since then much carbon monoxide (CO), carbon dioxide (CO_2) are generated more and more, and oxygen (O_2) reduced less and less, so the safe escape time i.e. allowed escape time should be within 5~8 minutes.

For public buildings and high-rise buildings of first or second fire resistance rating(FRR) the allowed escape time is 5min~7min;

For common buildings of third or fourth FRR, the allowed escape time, is 2min~4min;

For cinemas and theatres of first or second FRR, the allowed escape time 2min; of third FRR, 1.5min;

For gyms(gymnasiums)of first or second FRR, 3min~4min.

(7) Allowed maximum escape distance (AMED)See Table 2-5a,2-5b

Table 2-5a Allowed Max. Escape Distance(m)
允许的最大逃生距离

Types of Buildings 建筑类型	Un-porched Walkway 非袋形走道 FRR			Porched Walkway 袋形走道 FRR		
	1st or 2nd	3rd	4th	1st or 2nd	3rd	4th
Nursery 托儿所 Kindergarten 幼儿园	25	20	—	20	15	—
Hospital, sanatorium (Sanitarium, rest home) 医院、疗养院	35	30	—	20	15	—
School 学校	35	30	—	22	20	—
Other civil buildings 其他民用建筑	40	35	25	22	20	25

表 2-5b AMED of High-rise Buildings(m)
高层建筑允许的最大逃生距离(m)

	Un-porched Walkway 非袋形走道	Porched Walkway 袋形走道
Hospital's sickroom (ward) 医院病房	24	12
Hospital's other rooms 医院其他房间	30	15
Buildings of education, hotel and museum 教育建筑、旅馆和博物馆	30	15
Other buildings 其他建筑	40	20

室外防火距离见表 2-6 | Outdoor fireproof distance see Table 2-6

表 2-6 室外防火距离(m)
Outdoor fireproof distance(m)

High-rise building(HB) 高层建筑	HB 高层建筑	Podium 裙房	Other civil buildings 其他民用建筑		
			1st or 2nd FRR	3rd FRR	4th FRR
	13	9	9	11	14
Podium 裙房	9	6	6	7	9

2.10 屋顶

(1)屋顶的功能

为何要屋顶？

1)传统观念

屋顶应能防风、雨、雪、雹、尘、尘暴、雷击、火、地震、内爆、外炸、冬冷、夏热和直晒，此外屋顶还有分隔内外空间的功能。

2)可持续发展的概念

屋顶除上述功能外还应是多能转换器，将太阳能由集热器转换成热能提供热水；由太阳能电池转换成电提供照明与动力；由屋顶花草植物转换成光合作用能等。

2.10 Roofs

(1)Roof Functions

Why do we need roof?

1)Conventional concept

Roof must protect occupants against wind, rain, snow, hailstone, dust, dust storm, lightning, fire, earthquake, indoor explosion, outdoor bombing, winter cold, summer hot and insolation, besides, it divides indoor space from outdoor space.

2)Sustainable development concept

Besides the above, roof should be a multi-function convertor to convert solar energy into heat by solar collector to supply hot water; into electricity by solar cells to supply electric lighting and power; into photosynthetic energy by roofgarden plants, etc.

(2) 屋顶类型

见图 2-50, 2-51, 2-52, 2-53。

彩图 No. 3, 4, 5, 7, 8, 10, 13, 15, 16, 26, 27, 29, 38, 46。

我们将重点讨论 RC 平顶。

(2) Roof Types

See Fig. 2-50, 2-51, 2-52, 2-53。

Color pictures No. 3, 4, 5, 7, 8, 10, 13, 15, 16, 26, 27, 29, 38, 46.

We'll focus on the reinforced concrete (RC) flat roof.

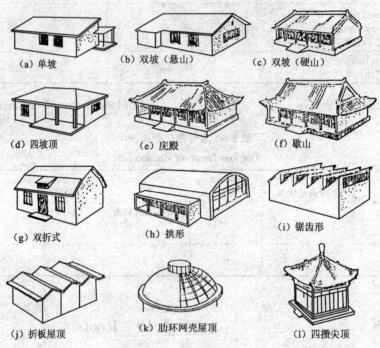

Fig. 2-50 Common roof types
图 2-50 一般屋顶类型
(a) shed roof(mono-pitched roof, single pitch roof, pent roof, half-span roof);
(b) xuanshan roof; (c) yingshan roof; (d) hip roof; (e) wudian roof; (f) xieshan roof;
(g) double folded roof; (h) arched roof; (i) saw-tooth roof; (j) folded-plate roof;
(k) net(web)-shell roof with surrounding rib; (l) pyramid roof

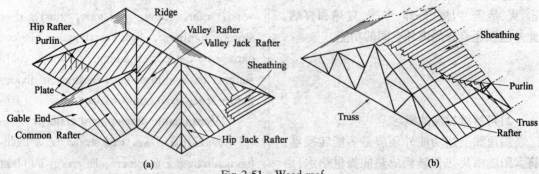

Fig. 2-51 Wood roof
图 2-51 木房顶

Ridge 正脊; Hip Rafter 斜脊; Sheathing 屋面板(望板); Purlin 檩子; Truss 屋架; Valley Rafter 斜沟主椽;
Valley Jack Rafter 斜沟支椽; Hip Jack Rafter 斜脊支椽; Common Rafter 普通椽; Gable End 端山墙;
(a) Frame Construction Roof; (b) Roof with Trusses, Purlins, Rafters, and Sheathing
(a) 木结构屋顶; (b) 木房顶的屋架、檩、椽、望板布置情况

2　建筑科学基础　　97

Fig.2-52　single-layer cable-supported roof
图 2-52　单层悬索屋顶

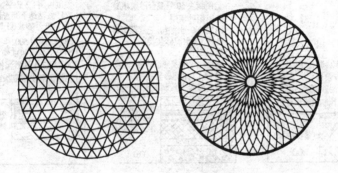

Fig.2-53(a)　A vertical view of the lamella wood net
structure dome roof's patterns and associates
图 2-53(a)　圆顶薄木网架俯视图案与交接

Fig.2-53(b)　An example of a lamella wood net structure dome roof
图 2-53(b)　圆顶薄木网架结构例子

(3) 钢筋混凝土平顶（坡度 $i \leqslant 10\%$）

屋顶构造有繁简之分，RC 平顶是最广泛采用的一种形式，它由三个基本部分组成：承重、绝热、防水层。旧法防水二毡三油绿豆砂（图 2-54(a)）。热毒施工缺点大，污气、伤人不环保，新法取代旧法是必然趋势。

(3) RC Flat Roof (Slope $i \leqslant 10\%$)

Roof construction can be complex or simple, RC flat roofs are widely used. Its basic components are build-up by three layers: bearing deck, insulation and waterproofing. In old method, waterproofing includes 2-ply bitumen felts, 3-course mastics and one course gravel (mineral granule, pea grit) (Fig.2-54a). Construction under very hot and poisonous conditions which is a serious shortcoming polluting air, damaging workers and environment. New methods arise instead of the old that's certain.

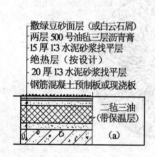

Fig.2-54　The old built-up RC flat roof

图 2-54　老法施工 RC 平屋顶

新法中，新卷材、新涂料，施工时常温、无毒保安全。

上人屋顶应加硬质铺面，空铺更比实铺保温隔热效果好（图 2-54(b)、(c)）。具有多能转换和屋面种植功能的屋顶应大力发展。

In new methods exist new rolls and new coatings, safety is sure in construction owing to normal temperature and innocuity.

A RC flat roof on which people can act must have a hard cover. An air space under the cover is better for insulation than that one without the air space (Fig.2-54(b),(c)). The roof which has the function of multi-energy conversion and roofing plants should be greatly developed

(4) 排水

排水方式有两种，即有水落管与无水落管（自由落水）（图 2-55），有水落管排水屋面设计见图 2-56。

(4) Drainage

Drainage has two forms: piped and without pipe (free fall) (Fig.2-55), design of piped drainage see Fig.2-56.

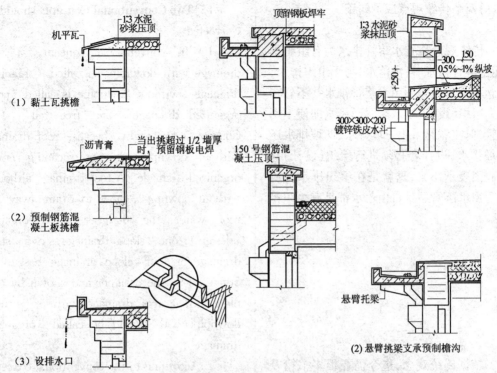

Fig.2-55 Roof drainage: by downpipe or free fall

图 2-55 屋面排水：有水落管或自由落水

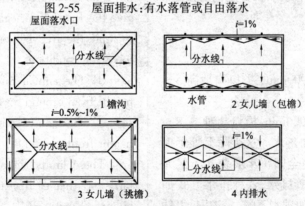

Fig.2-56 Roof piped drainage design

图 2-56 屋顶有水落管排水设计

自由排水好

自由落水的优点是组件简、施工易、排水畅、渗漏少、造价低、管理工作少。自由落水的优点恰好就是有水落管排水的缺点。

Free fall drainage good

The advantages of free fall drainage are simple components, easy construction, less stop, low escape, cheap cost and minimum managements.

The advantages of free fall drainage are just the shortcomings of the drainage by downpipe.

(5) 两个传统观点应予纠正

1) 传统观念称有水落管排水为有组织排水，无水落管排水（自由落水）为无组织排水，此观念不妥。因为不用水落管排水并不意味着不去组织（设计）排水。屋面排水即使不用水落管，但究竟是采用一坡排水、两坡排水还是四坡排水，设计者必须进行组织（设计）并用屋面图表示出来，这就是在组织排水。所以不管用水落管还是自由落水都是有组织排水。

2) 依据传统观念，迄今仍在很多书籍及设计文件中对自由落水采取否定的态度，如规定：

①年降雨量＞900mm，檐口离地高 5～8m；

②年降雨量＜900mm，檐口离地高 8～10m。方可采用自由落水。这实际上等于扼杀了自由落水，并使全国每年浪费几十亿元人民币用于有水落管排水，这种排水方式如前所述，缺点甚多，如排水不畅、易渗漏，组件多，施工难，造价高，管理工作多。

为什么传统观念对自由落水持否定态度呢？据调查，有下列误解，传统观念认为：

(5) Two Conventional Concepts should be Corrected

1) In conventional concept, a roof drainage with downpipe is called "organized drainage", without downpipe is called "non-organized drainage" or "free fall". This Concept is not right, because roof drainage without downpipe is not meaning non-organized drainage. A roof drainage although without downpipe, how to drain away the roof water the designer must organize (design): one-slope drainage, two-slope drainage or four-slope drainage has to be organized by the designer and shown by roof plan. So, a roof drainage with or without downpipe both should be called "organized drainage".

2) According to the conventional concept, up to now many books and design documents give the free fall a negative attitude, such as to limit the free fall only being used in the following conditions:

① The annual rainfall ＞ 900mm, eaves height from ground 5～8m;

② The annual rainfall ＜ 900mm, eaves height from ground 8～10m.

These limits actually throttle the "free fall" and that each year would waste several billion Yuan RMB in using downpipe drainage. As above explained; it has many shortcomings: drainage not fluent, easy escape, many components, difficult construction, high cost and more management works.

Why the conventional concept gives the free fall a negation attitude? Based on investigation, there exist some mistakes. In conventional concepts:

①自由落水易受风吹击墙面并污染墙面;

②自由落水会损伤散水;

③自由落水自散水面回溅水污染勒脚墙;

④滴水掉头上,使人难受;

⑤冷区,檐口冰柱砸伤行人。

作者本人也曾长期存在上述错误观念,但是在通过实际观察、科学试验,请教有关专家,重新学习流体力学后,证实原来的错误观念乃是一种主观意念,是没有科学依据的,这才得到了彻底的纠正。

为什么檐口水不会撞墙?

①吹纸试验　如图 2-57,手指夹一张薄信纸于口部下方,纸自然下垂。口向前吹风,纸会向上飘起,这是什么原因呢?

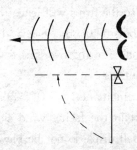

图 2-57　吹纸试验

根据流体力学原理,未吹风时,纸面前后受到的静压能是相等的。口吹风时,纸面前空气产生流动,使原静压能变为动能,因而静压能小于纸面后方的静压能,于是纸就被抬(吸)起来了。

①Free fall drips under wind easily impact and pollute wall;

② Free fall drips easily damage the apron;

③ The rebound water from the apron pollutes the plinth wall;

④Dripping eaves water on head causes uncomfort;

⑤In cold region, eaves ice sticks impact walkers.

The author-self in a long period also had the above mistake concepts. Through practical observation, science experiments, asking relative specialists and relearn the hydromechanics which have proved that in my mind, the original wrong concepts were a subjective thought without any science basis and then got the whole correction.

Why Eaves Drops never impact walls?

①Paper blew test

See Fig. 2-57. Clip a thin letter paper with fingers under your mouth, it will be naturally hung there. Blow the air to front with your mouth, the paper will fly upwards. Why?

Fig. 2-57　Paper blew test

According to the hydromechanics principle, when no blowing, the static pressure energies on both sides of the paper are equal, while blowing, in front of the paper the air is moving which makes the original static pressure energy dynamic energy causing the static pressure energy reduced and the paper is lifted (absorbed) to fly upwards.

结论：在流体运动中，物体必向流速大的一侧移动。

观察实际现象与试验均证实：风吹向墙面时，檐口自由落水不仅不会被风吹击墙面，反而会偏离墙面。图 2-58a)为风吹过建筑物的动态图。迎风面，接近墙面处，风将转向，变得愈近墙面气流愈平行于墙面，这表明垂直于墙面的风速越近墙面越小，最后为零。风向平行于墙面区域称边界区如图 2-58a)2-2 至 3-3 区。①檐口自由落水在边界区内下落将受不到垂直吹向墙面的风推力作用；②图 2-58b)表示檐口下落水滴外侧流速 V_0 大于其内侧流速 V_i，故水滴将向外（偏离墙面）移动；③檐口水在脱离檐口下落的瞬间还会受到气流向上、向外绕檐而过的外推作用。①、②、③就是迎风面檐口自由落水不仅不会吹击墙面，反而会偏离墙面的原因。

我们多次试验也证实了上述现象。

背风面如图 2-58(a)所示，檐口水处于负压区，将被吸离墙面。

②我们曾观察使用 20 多年以上的混凝土散水，三层楼高自由落水撞击下有一条浅色滴水带区，个别处有 1~1.5mm 的深痕，凡

Conclusion: In a fluid motion, a body certainly moves to the fast side.

Observation of practical phenomena and experiments have proved that when wind blows to the wall, the eaves free fall drips are not blew to impact the wall, inversely they will fly to leave the wall. Fig. 2-58(a) is a dynamic showing of wind flying over a building. In facing wind side, near the wall, wind will change its direction, the more near the wall, the wind runs more parallel to the wall which explains that the wind speed vertical to the wall will be smaller and smaller while near and near the wall, at last be zero. The area of wind parallel to the wall called "bound area" as shown in Fig. 2-58 (a) 2-2 to 3-3 range. ① When eaves drips (drops) fall in the bound area, there is no any wind force vertical to the wall to put the drips; ② Fig. 2-58(b) shows the wind speed V_0 outside the drips is faster than the wind speed V_i inside the drips, so the drips will leave the wall; ③ At the moment of the eaves drips falling down, the air flowing over the eaves will put the drips outside.

The combination of ①、② and ③ can answer that in the facing wind side why the eaves free fall drips do not be blew to impact the wall, inversely, are flying to leave the wall. Our many times experiments have also proved the above phenomenon.

In the back of the building, as Fig. 2-58a) showing, eaves drips being in the negative pressure area will be absorbed to leave the wall.

②We have observed a concrete apron having been used for more than 20 years, under the 3-story building eaves drips impacting, there exists a

卵石面上均无任何撞痕。可见混凝土散水的耐久性是很足够的。

light color tie area, several impacted marks with 1~1.5mm depth distributed in the tie area. On the pebble surfaces, no any impacted marks discovered. The above has proved enough that the concrete apron's durability is very strong under eaves drips impacting.

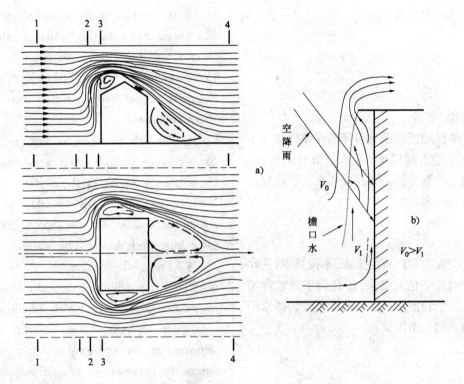

Fig. 2-58 The dynamics as wind over a building and the eaves drips falling down

图 2-58 风绕房屋及檐口水下落动态

③我们曾观测多处多层建筑物自由落水,并用自来水(自三楼冲击散水相当于几十层高楼滴水冲击)试验,散水面回溅水高均不超过 65cm,故从保护勒脚墙看,勒脚抹灰 70cm 高已足够了。

③We have observed many buildings in many places the eaves free fall drips and had an experiment with tap water from a 3-story building impacting the apron (it's alike to the eaves drop impacting from a several tens stories high-rise building), all the resilient water does not be higher than 65cm. To protect the plinth wall, a waterproof finish of 70cm height is enough.

④散水不是人行道,雨天有人偶在散水上行走受自由落水滴湿乃是偶然的现象,不能作为否定自由落水的科学依据。

④An apron is not a pavement, in a raining day somebody occasionally walking on the apron being wet by the eaves drips which would be a

⑤冷区冬季檐口冰柱砸人事件,迄今未闻一实例(即使有,那也是极偶然的事)。冰柱只会在暖季来临时在阳光及暖气流作用下融化成水滴滴落,不会整截冰柱或冰块掉落除非有撞击力作用。

结论

①淋湿墙面的是天空降雨在惯性作用下击穿边界层击湿墙面,不是檐口自由落水。怕檐口水淋湿并污染墙面设立水落管收集檐口水实是一大浪费。

②应大力推广自由排水,不仅效率好,每年可节约几十亿元资金,还有利于散水种植并消除湿陷和湿胀地区水落管不均匀排水引起房屋不均匀下沉。

③出入口处不用自由落水。

④屋面、阳台、雨棚等建筑构件排水无论有管或无管都是有组织排水。中国古典建筑如歇山、庑殿屋顶都没有水落管,但却是井井有条的有组织排水。

2.11 建筑夏季防热

图2-59(a)所示为建筑物夏季得热途径。

⑤ We never have heard that in cold region cold seasons, eaves ice sticks damaged people(If anywhere might be happened that was only a happenchance). When warm season is coming, under sunshine and warm air, the ice sticks start to be melted into water which then drops down in point by point. However, no any ice bars or blocks drop down except an impact force to impact them.

Conclusion

①The wall is wet by the sky rain, not by the eaves drips. The bound area is staved in by the sky rain depending on the inertia of the rain. We have been afraid of the eaves drips to wet and pollute the wall so set downpipe to collect the eaves drips which does be a big waste.

② Free fall drainage should be greatly developed, high efficiency, annual cost saving of several tens billion Yuan RMB, benifiting apron plant and canceling the building unequal settlement in wet-settlement and wet-expansion regions due to downpipes' unequal drainage.

③Free fall drainage should not be used at entrance.

④ The drainages of roof, balcony, canopy,etc. are all organized drainage. Chinse ancient buildings such as xieshan roof, wudian roof, are all free fall drainage without any downpipe but they all are organized drainage in good order with line by line.

2.11 Summer Heat Insulation of Buildings

Fig. 2-59(a) shows summer heat gain ways in a building.

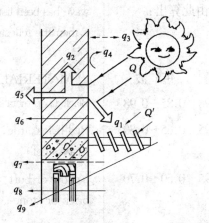

Fig.2-59(a)　In a building summer heat gain ways
图 2-59(a)　建筑夏季得热途径

Q: Solar radiation casted on wall/roof;
q_1: Reflect heat; q_2: Absorbed heat;
q_3: Radiant heat exchange between the building and suroundings;
q_4: Convective heat exchange;
q_5: Transmitted heat; q_6: Heat gain by air temp. differential between indoor and outdoor;
q_7: Heat gain through cracks;
q_8: Heat gain by air temp. differential between both sides of window;
q_9: Transmitted heat through window;
Q': Solar radiation casted on window.

Q: 投射到屋顶和墙上的太阳辐射热；
q_1: 反射热；q_2: 吸收热；
q_3: 建筑物与周围环境辐射换热；
q_4: 对流换热；
q_5: 透过热；q_6: 室内外温差得热；
q_7: 隙缝进热；
q_8: 窗两侧温差得热；
q_9: 通过窗户的透射热；
Q': 投射到窗户上的太阳辐射热。

如何减少夏季得热？

1) 反射

思考：

①白色与铝箔，哪个反射太阳辐射热好？

②用反射防钢厂热钢锭的辐射，白色与铝箔那个好？

参见下表找答案。

建筑材料吸收率（α）和发射率（ε）

表中分别为长波和短波的 α 值，在同样

How to reduce summer heat gain?

1) Reflection

Thinking deep:

①Under solar radiation, which is better to reflect the radiation, white color or aluminium?

②In steel mill to reflect the hot steel ingot(热钢锭) radiation, white color and aluminium which is better?

See the following table to find the answer.

ABSORPTANCES (α) & EMITTANCES (ε) OF BUILDING MATERIALS

In the table the "α" of long wave and short

波长时，$\alpha = \varepsilon$，而反射率 ρ 可由此算出：
$$\rho = 1 - \alpha。$$

wave has been listed, at the same wave length, $\alpha = \varepsilon$, then the reflectance ρ may be computed by:
$$\rho = 1 - \alpha。$$

材料	长波	短波	MATERIAL	Long Wave	Short Wave
黑色表面(非金属)	0.90~0.98	0.85~0.98	Black (non-metallic)	0.90~0.98	0.85~0.98
红砖,瓦,锈钢	0.85~0.95	0.65~0.80	Red brick, tile, rusty steel	0.85~0.95	0.65~0.80
黄或浅黄砖、石	0.85~0.95	0.50~0.70	Yellow or buff brick or stone	0.85~0.95	0.50~0.70
米色砖,浅色	0.85~0.95	0.30~0.50	Cream birck, light colours	0.85~0.95	0.30~0.50
发亮白涂料	0.85~0.95	0.15~0.20	White(bright) whitewash	0.85~0.95	0.15~0.20
发亮铝漆	0.40~0.60	0.20~0.30	Aluminium paint(bright)	0.40~0.60	0.20~0.30
发暗铝金属、镀锌钢材	0.20~0.30	0.40~0.65	Aluminium metal(dull), galvanised steel	0.20~0.30	0.40~0.65
高抛光金属铬、铝	0.20~0.40	0.05~0.15	Metal (highly polished) chrome, aluminium	0.20~0.40	0.05~0.15

冷藏库屋顶与外墙外表面用什么材料做终饰好？

思考：反射是可持续方法吗？

2)通风

图 2-59b)为通风屋顶几个剖面,图 2-60 为一砖拱通风屋顶实例,华南广州市。

3)阻存

图 2-61 为用屋顶水池、屋顶种植和中国窑洞为阻存物质减少夏季室内得热。

What material is suitable to be the outer finish of the roof and exterior wall of cold storage?

Thinking deep: Is the reflection a sustainable method?

2) Ventilation

Fig. 2-59(b) shows some sections of roof ventilation, Fig 2-60, a practical example of brick arch ventilation roof in Guangzou City, Southern China.

3) Resistance & storage

Fig. 2-61 shows the utilization of roof pool, roof plant and Chinese cave as the resistance & storage mass (thermal mass) to

4）遮阳

reduce the summer heat gain in buildings.
4) Shading

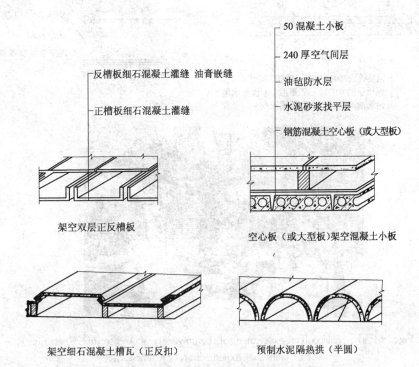

Fig.2-59(b)　Roof ventilation sections
图 2-59(b)　通风屋顶剖面

Fig.2-60　Brick arch ventilation roof, Guangzhou
图 2-60　广州砖拱通风屋顶

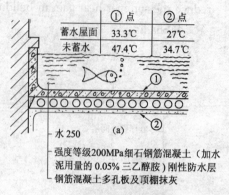
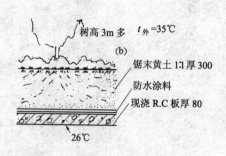

Fig.2-61(a) roof pool(an experiment of the university of Archtecture, Chongqing), (b)roof plant(our experiment), (c)Chinese cave

图 2-61(a) 屋顶水池(重庆建大实验)(b)屋顶种植(我们的实验)(c)中国窑洞

不同窗户系统遮阳系数

遮阳系数(SC)的定义是,通过遮阳窗与无遮阳(DS)3mm 透明玻璃窗太阳进热之比

系　统	遮阳系数
标准 3mm 玻璃窗	1.00
内侧(取决于颜色与遮挡范围)	
深色可卷百叶半关到全关	0.80~0.90
中等颜色可卷百叶半关到全关	0.60~0.7
浅色软百叶帘半关到全关	0.45~0.55
铝箔反射软百叶半关到全关	0.4~0.5
浅色织布帘	0.4

SHADING COEFFICIENTS FOR VIRIOUS WINDOW SYSTEMS

System	Shading Coefficient
Regular 3mm window glass	1.00
Inside (depends on colour and extent of closure)	
Dark coloured roller blinds half to fully closed	0.80~0.90
Medium coloured roller blinds half to fully closed	0.60~0.7
Light coloured venetian blinds half to fully closed	0.45~0.55
Reflective aluminium venetian blinds half to fully closed	0.4~0.5

浅色可卷百叶全关	0.4	Light coloured fabric curtain	0.4
		Light coloured roller blind fully closed	0.4
玻璃(细节见厂方资料与详细说明)		Glass(for details see manufacturers literature and specifications).	
单层6mm透明平板玻璃	0.95	Single 6mm clear float	0.95
吸热玻璃	0.45~0.7	Heat absorbing glasses	0.45~0.7
热反射玻璃	0.25~0.5	Heat reflecting glasses	0.25~0.5
双层6mm透明平板玻璃	0.83	Double 6mm clear float	0.83
吸热玻璃	0.4~0.6	Heat absorbing glasses	0.4~0.6
反射玻璃	0.15~0.4	Heat reflecting glasses	0.15~0.4
外侧(取决于颜色与遮阳范围)		Outside(depends on colour and extent of closure)	
遮阳树,不浓厚	0.5~0.6	Shade tree, light	0.5~0.6
遮阳树,浓厚	0.2~0.25	Shade tree, heavy	0.2~0.25
外侧水平或垂直百叶	0.1~0.2	Outside louvres, horizontal or vertical	0.1~0.2
外侧浅色软百叶	0.1~0.2	Outside light coloured venetian blinds	0.1~0.2

5)绿化

特别是季青植物夏季叶浓可遮阳,冬季叶落,室内可得到温暖的太阳热。

6)多能构件转化太阳能为热、电、绿化能等。

7)自然空调

例如利用地下通风道调节入室空气的温、湿度。

8)优化组合上述措施。

2.12 门窗

(1)门类型

门的功能:日常交通;紧急疏散;分隔建筑空间;其他围护功能如绝热、隔声、防火等。

5)Greening

Especially seasonal green plants have shock leaves in summer to shade the solar radiation, in winter, leaves fall,indoor can get warm solar heat.

6)Multi-function members convert solar energy into heat, electricity and greening energies,etc.

7) Natural air conditioning, such as use an underground ventilation tunnel to adjust the coming air temperature and humidity.

8)Optimal composite of the above.

2.12 Doors and Windows

(1)Door Types

Door functions: daily traffic; urgent escape; isolating building spaces; other protective functions, such as, heat insulation/sound insulation and fireproof.

110　建筑科学基础

门类型见图 2-62

平开门得到了广泛的应用。图 2-63 为平开门的构造。

图 2-64 为保温、隔声门框—扇连接密闭构造。

Door types see Fig.2-62.

The widely used doors are the swinging doors (平开门)。Fig. 2-63 shows a wood swinging door's construction.

Fig.2-64, constructions of framing-sash joint seals of heat & sound insulation doors.

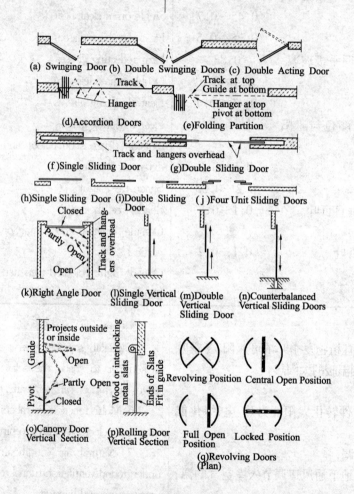

Fig.2-62　Door types

图 2-62　门类型

(a)单扇平开门；(b)双页平开门；(c)弹簧门；(d)中悬折叠推拉门(Hangen 滑轮挂钩.Track(轨道)；(e)边悬折叠推拉门(滑轮挂钩与轨道在顶部，导向与枢轴在底部)；(f)单扇滑门(可单向滑入墙内)；(g)双扇滑门(可双向滑入墙内)；(h)单扇贴墙滑门；(i)双扇贴墙滑门；(j)4 扇贴墙滑门；(k)直角开关门；(l)单扇垂直滑门；(m)双扇垂直同向滑门；(n)双扇垂直异向滑门；(o)折叠上开门；(p)卷帘门；(q)旋转门

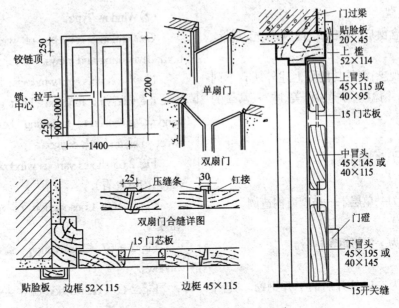

Fig. 2-63 Wood swinging door construction
图 2-63 木平开门构造

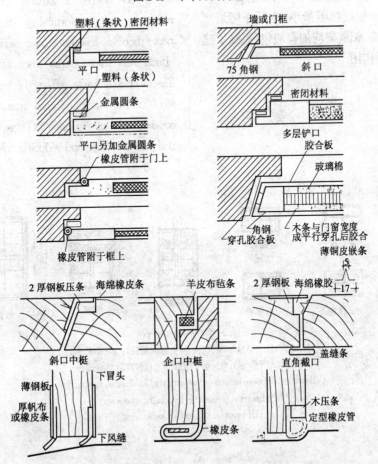

Fig. 2-64 Constructions of framing-sash(casement)joint seals of heat & sound insulation doors
图 2-64 保温(隔热)、隔声门框与扇接缝密封构造

(2) 窗类型

传统观念窗户的功能是:通风、采光、观景。

新窗户观念,窗的功能除上述外还有下列功能:冬得热,窗温室可提供色美、味香、富氧的空间。

图 2-65 为各类窗例。

图 2-66 为一单层木—玻璃窗构造例。

图 2-67 为一实腹钢窗构造。

图 2-68 为一向外开启的平开钢窗。其滴水槽好吗?

木窗框热阻大、舒适性好。钢窗框热阻差,冷桥作用严重。铝框热阻最小,冷桥作用最严重。当今钢框架、铝框架均加有塑包,提高了热阻、减低了冷桥作用。

(2) Window Types

Conventional concepts of windows: ventilation, daylighting and views.

New concepts of windows: Besides above, have the following functions: Heat gain in winter, window greenhouse supplying beautiful colour, sweet smell and rich oxygon.

Fig.2-65 shows various window types. (inside views, 内→外看).

Fig.2-66 Construction of a wood-glass window.

Fig.2-67 Construction of a solid steel windows.

Fig.2-68 Construction of a steel outswinging casement. The drip groove is good?

Wood frames of window have good thermal resistance and comfortability. Steel frames have poor thermal resistance and serious cold bridge effect. Aluminium frams' thermal resistance is the least and cold bridge effect the most. Now either steel frames or aluminium frams all have been coverd with an envelope of plastics to increasing thermal resistance and reducing cold bridge effect.

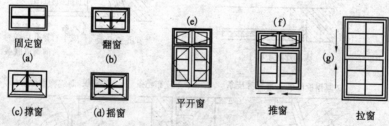

Fig.2-65 Window types
(a)固定窗;(b)(下悬外翻窗);(c)(上悬内翻窗);(d)(中悬窗);
(e)(向内平开窗);(f)(水平推拉窗);(g)(双扇悬滑窗)
(a)Fixed; (b)Bottom-hanged outswinging; (c)Top-hanged inswinging;
(d)Horizontal pivoted; (e)Inswinging casement;
(f)Horizontal siding; (g)Double hung

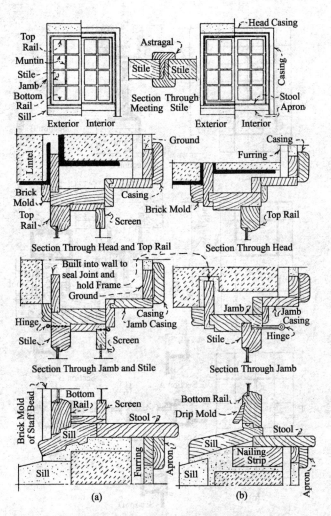

Fig. 2-66 Wood-glazing window construction
图 2-66 木——玻璃窗构造

Top Rail 上冒头;Muntin 窗格条(棂子);Stile 边梃;Jamb 边框(槛);
Bottom Rail 下冒头;Astragal 密封盖条(开关缝处);Lintel 过梁;Brick Mold 贴脸板;Casing 罩板(窗套);
Ground 底板;Built into wall to seal joint and hold frame 嵌入墙体以便封闭接缝并固定窗框;Screen 纱窗;
Hinge 铰链;Stool 内窗台;Apron 内窗台下护盖;Staff Bead 纤维起浆凸缘线脚;Sill 外窗台
(a)Casement Window Opening Out 向外平开窗;(b)Casement Window Opening In 向内平开窗。

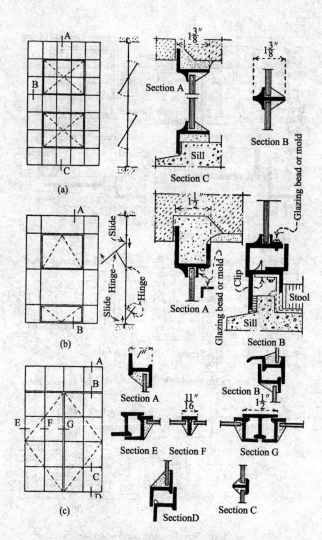

Fig.2-67 Solid-section steel windows

图 2-67 实腹钢窗剖面

(a)Horizontally Pivoted Window 中悬窗;
(b)Projected Window,Architectural 建筑用撑滑窗;
(c)Side-Hinged Casement Window 边铰链平开窗;

Slide 滑动方向;Hinge 铰链;Glazing bead or mold 压玻璃条

(其余见图 2-66 汉—英对照词)

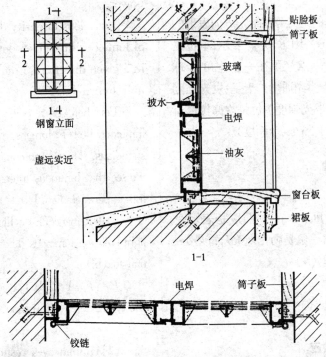

Fig. 2-68 The construction of a outswinging steel window
图 2-68 向外平开钢窗构造

2.13 天然采光

思考:夜,皓月当空,有光否?

光是眼——脑系统对一定范围电磁波的主观反应。

(1)测光单位

1)发光强度 早期发光强度称烛光 candel power,一支标准的鲸油蜡烛光在观测方向产生的光强度叫做烛光 candel power(candel)。现用发光强度单位叫坎(德拉)candela,1 candela (cd) = 0.981 candel power.

2.13 Daylighting (Natural Lighting)

Thinking deep: At night, the moon shines brigtly, is there any light?

Light is the subjective response by the eye - brain system to a certain range of electromagnatic waves.

(1) Photometric Units

1) Luminous Intensity The early luminous intensity was called candel power. A standard sperm wax candel was described together with a direction of observation, the resulting intensity being called the candel power.

Today the luminous intensity is called candela, 1 candela (cd) = 0.981 candel

发光强度表示的是光源在给定方向发射的光通量(luminous flux),用 F 或 Φ 表示,单位是流明(符号 lm)。设一点光源在一立体角 ω 球面度内发射出 F(流明光通量),于是:发光强度 $I=F/\omega$,由此可知发光强度单位是坎德拉或流明每球面度,1 cd=1 lm/sr(球面度)。

2)照度 (E) Illuminance。照度是表示受光面受到的光通量,单位是勒克斯 lux,符号 lx.
$E=F/A$(受光面积)(lx)

(2)人眼感光效应

图 2-69 描述了人眼的感光效应。

从图 2-69 可以看出人眼感光效应:

1)光感:从该电磁辐射中,在波长 380nm 到 730nm 间人的眼——脑系统能转换成光(感),称"可见光"。(另有数据:380nm～760nm)。

2)彩色感:人的眼——脑系统能将不同波长的光转换成不同的彩色感如图 2-69 所示。最通常的划分就是牛顿的划分法,即:红、橙、黄、绿、蓝、青、紫。

3)人眼色彩感灵敏度与色彩叠加效应

明视条件即在明亮环境人眼最灵敏感是波长 555nm 的黄绿色。

power.

Luminous intensity describes the quantity of luminous flux given out by the light source in a given direction. Luminous flux(F or Φ), unit:lumen(lm).

Let there be a point light source emitting luminous flux F lumens in a solid angle of ω steradians. Then:Luminous intensity $I=F/\omega$, so we see that luminous intensity unit is candela or lumen per steradian. 1 cd=1 lm/sr.

2)Illuminance(E) Illuminance describes the light emitting luminous flux F lumens on an area, unit lux(lx).

$E=F/A$ (lx), it can be measured by the Lux Meter(Illuminometer,Lightmeter).

(2) Human eyes respondent effect to the Light

Fig. 2-69 shows the respondent effects of human eyes to the light.

From Fig. 2-69 we can see the effects of human eyes' response to the light:

1) Light perception: in the electromagnetic radiation, between the wave lengths of 380 nm to 730nm the eye-brain system is able to convert into light called "visible spectrum"(可见光)(Another data:380nm～760nm).

2)Colour perception:the eye-brain system can convert different wave length lights into different colours as showing in Fig. 2-69, the most common being those allocated(划分) to the colours by Newton i.e.red, orange, yellow, green, blue, indigo and violet.

3) Color sensitivity and piled effect of human eyes

Light adapted i. e. photopic condition, the most sensitivity of human eyes is at 555nm

暗视条件即微光环境人眼最敏感是波长507nm 的绿蓝色。详见图 2-69。

图 2-70 示出了不同色彩在白底上叠加后人眼的视觉效应。

Dark adapted i. e. scotopic condition, at 507nm green-blue. Detail see Fig. 2-69.

Fig. 2 - 70 has shown different color piled on a white base the effective perception of human eyes.

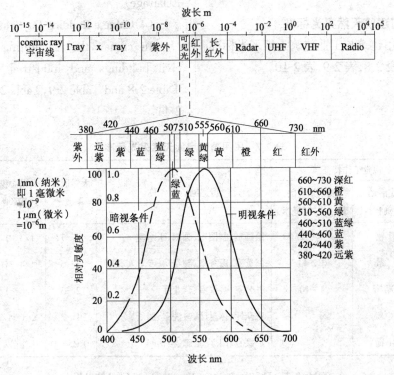

Fig. 2-69　A diagram of being eyes' respondent effect to the light

图 2-69　人眼感光效应图（以平均眼为准）

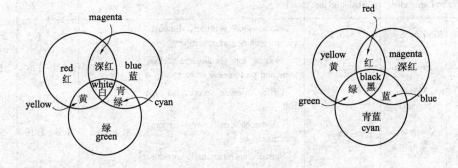

Fig. 2-70　Different color piled on a white base the effective perception of human eyes

图 2-70　不同彩色叠加在白底上的视觉效应

(3) 天然采光光源标准

天然采光理论上是利用国际照明委员会标准全阴天天空光为准。

(4) 中国建筑天然采光标准

中国民用建筑与工业建筑天然采光标准见表2-7与表2-8,表2-9,表2-10。

(3) The Standard of Daylighting Light Source

In the daylighting theory the utilization of sky light of the CIE Standard Overcast Sky is the basis.

CIE (Commision International del'Eclairage)

(4) Chinese Building Daylighting Standards

In china, the daylighting standards of civil buildings and industrial buildings see Table 2-8 and Table 2-9, Table 2-10 and Table 2-11.

表 2-7 民用建筑房间天然采光标准

采光等级	视觉工作特征		房间名称	天然照度系数[①]	采光面积比 玻/地	天然照度 lx
	工作或活动要求精确程度	要求识别的最小尺寸(毫米)				
I	极精密	<0.2	绘画室、制图室、画廊、手术室	5~7	1/3~1/5	250~350
II	精密	0.2~1	阅览室、医务室、健身房、专业实验室	3~5	1/4~1/6	150~250
III	中等精密	1~10	办公室、会议室、营业厅	2~3	1/6~1/8	100~150
IV	粗糙		观众厅、休息厅、盥洗室、厕所	1~2	1/8~1/10	50~100
V	极粗糙		贮藏室、门厅、走廊、楼梯间	0.25~1	1/10以下	25~50

Table 2-7 The Daylighting Standards of Civil Buildings

Daylighting Ratings	Character of Eyemark		Rooms	Natural Illumination Factor(NIF) (%)	Rate of Glazing Area/Floor	Natural Illumination (lx)
	Accuracy Demanded	Min size of Distinguishability(mm)				
I	Most precision	<0.2	Rooms for painting, drawing, picture gallery and operating	5~7	1/3~1/5	250~350
II	Precision	0.2~1	Rooms for reading, medicine, gym, and professional laboratory	3~5	1/4~1/6	150~250
III	Middle precision	1~10	Office, meeting room, business hall	2~3	1/6~1/8	100~150
IV	Coarseness		Audience hall, rest room, washroom, wc	1~2	1/8~1/10	50~500
V	Very coarseness		Storage, entrance hall, corridor, staircase	0.25~1	<1/10	25~50

表2-8 生产车间和工作场所的采光等级举例
Table 2-8 Examples of Daylighting Rating of Work-shop and Work-site

采光等级 Daylighting Ratings	生产和工作状况 Producing and Working Situation
Ⅰ	精密加工 precise,检验 check,雕刻 carving,刺绣 embroidering,绘画 drawing(painting)
Ⅱ	装配 assembly,主控制室 main control room,检修 overhaul,光学元件加工 photo member processing,排字 composing,印刷 pressing
Ⅲ	机修 machine repair,一般控制室 general control room,理化实验室 physics and chemistry laboratory,计量室 measuring room,木工 woodworking,冷轧 cold rolling,热轧 hot-rolling,拉丝 wiredrawing,发电厂锅炉房 power station boiler room
Ⅳ	焊接 welding,钣金加工 sheet-metal processing,炼钢 steelworks,炼铁 ironworks,铸工 casting,电镀 electroplating,油漆 oilplant,化工厂 chemical,农药 agriculturaldrug
Ⅴ	汽车库 garage,锅炉房 boiler room,泵房 pump house,运输站 transportation station,一般仓库 general storehouse

表2-9 窗(玻)地面积比
Glazing Area/Floor Area

采光等级 Daylighting Ratings	天然照度系数最低值 Min Natural Illumination Factor(%)	单侧窗 One Side Windows	双侧窗 Double Sides Windows	矩形天窗 Rectangular Skylights	锯齿形天窗 Saw-toothed Skylights	平天窗 Flat skylights
Ⅰ	5	1/2.5	1/2.0	1/3.5	1/3	1/5
Ⅱ	3	1/2.5	1/2.5	1/3.5	1/3.5	1/7.5
Ⅲ	2	1/3.5	1/3.5	1/4	1/5	1/8
Ⅳ	1	1/6	1/6	1/8	1/10	1/15
Ⅴ	0.5	1/10	1/7	1/15	1/15	1/25

表2-10 生产车间工作面上的采光系数最低值

采光等级	视觉工作分类		室内天然光照度(lx)		天然照度系数(%)	
	工作精确度	识别对象的最小尺寸 d (毫米)	最低侧光	平均顶光	最低侧光	平均顶光
Ⅰ	特别精细工作	$d \leqslant 1.5$	250	350	5	7
Ⅱ	很精细工作	$0.15 < d \leqslant 0.3$	150	250	3	5
Ⅲ	精细工作	$0.3 < d \leqslant 1.0$	100	150	2	3
Ⅳ	一般工作	$1.0 < d \leqslant 5.0$	50	100	1	2
Ⅴ	粗糙工作及仓库	$d > 5.0$	25	50	0.5	1

注:1 采光系数最低值是根据室外临界照度为5000勒克斯制定的。如采用其他室外临界照度值,采光系数最低值应作相应的调整。采光系数即天然照度系数。
2 四川、贵州、广西地区的室外临界照度为4000勒克斯,表中各级的采光系数标准值应乘1.25的地区修正系数。

Table 2-10 Min Daylighting Illumination Factor at the Working Plane of Worksshop

Daylighting Ratings	Character of Eyemark		Natural Illumination (lx)		Natural Illumination Factor (%)	
	Accuracy Demanded	Min. Size of Distinguishability (mm)	Min side Lighting	Mean Roof Lighting	Min side Lighting	Mean Roof Lighting
I	Most precision	$d \leqslant 1.5$	250	350	5	7
II	Very precision	$0.15 < d \leqslant 0.3$	150	250	3	5
III	precision	$0.3 < d \leqslant 1.0$	100	150	2	3
IV	General work	$1.0 < d \leqslant 5.0$	50	100	1	2
V	Coarseness, store	$d > 5.0$	25	50	0.5	1

Note1: The min Daylighting Illumination Factorisdecided by the critical illumination (threshold illumination) 500lx. If using other outdoor critical illumination value the min. daylighting illumination factor should be correspondingly adjusted. The daylighting factor is just the natural illumination factor.

Note2: The outdoor critical illumination value of Sichuan, Guizhou and Guangxi is 4000 lx, the daylighting values as listed above should be multiplied by 1.25 the local correction factor.

(5) 天然照度系数

$C = \dfrac{E_i}{E_0} \times 100\%$ 如图 2-71 所示：

E_i：室内测点 lux (lx)；

E_0：室外同时测值 lux (lx)。

(5) Natural Illumination Factor

$C = \dfrac{E_i}{E_0} \times 100\%$ As Fig. 2-71 showing:

E_i: indoor measured point lux (lx);

E_0: at the same time outdoor lux (lx).

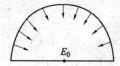 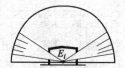

Fig. 2-71 Indoor and outdoor illumination measure
图 2-71 室内外照度测定

(6) 临界照度与天然照度系数最低值

(6) Critical Illumination (threshold illumination) and Minimum Natural Illumination Factor

$C_{\min} = \dfrac{E_{i,\min}}{E_{0,c.i}} \times 100\%$

$C_{\min}$：最低天然照度系数；

$E_{i,\min}$ 室内最低照度值(lx)；

$E_{0,c.i}$：室外临界照度值(lx)。

图 2-72 为不同窗户位置不同照度曲线。

$C_{\min} = \dfrac{E_{i,\min}}{E_{0,c.i}} \times 100\%$

$C_{\min}$: minimum natural illumination factor;

$E_{i,\min}$: indoor min. illumination (lx);

$E_{0,c.i}$: outdoor critical illumination (lx).

Fig. 2-72 shows different window positions different illuminance curves.

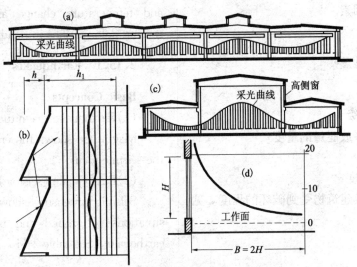

Fig. 2-72 Different window positions different lighting curves
图 2-72 不同窗户位置不同照度曲线
(a)Side & rectangular skylights;(b) Saw-toothed skylights;(c)High-side & side skylights;(d)Single side skylights
(a)侧窗与矩形天窗结合;(b)锯齿形天窗;(c)高侧窗与侧窗结合;(d)单侧窗

2.14 窗温室生态效益

将 RC 窗台做成 500～700mm 宽,内外用推拉玻璃窗扇,垫以 150mm 蛭石或珍珠岩作轻质种植层,种植矮株花卉即可构成窗温室。RC 窗过梁做成种植槽种植爬墙虎等植物,夏季叶密墙有遮阳,冬季叶落墙得日照。生态效益:(1)增强绝热隔声;(2)改善空气质量(吸 CO_2,吐 O_2);(3)提供美色、香味。

2.14 Eco-Benefit of window Greenhouse

Take the windowsill's width 500～700mm, and glass sliding windows set at inside and outside, on windowsill set a 150mm light material planting layer with vermiculite or pearlite (perlite), plant short flowers to form a window greenhouse. Make the RC window linter a plantable trough to plant Boston ivy, etc. which in summer with thick leaves can let the wall get shading, in winter leaves fallen, the wall get sunshine. Eco-benefits:(1)Increasing insulation of heat and sound;(2) Improving air quality (absorbing CO_2, releasing O_2);(3) Supplying beautiful color and sweet smell.

2.15 建筑物变形与抗变形

2.15 Deformation and Anti-deformation of Buildings

地震、湿陷、湿胀以及温度变化是引起建

Earthquake, wet-settlement, wet-expansion

筑变形的主要因素。

2.15.1 地震

基本概念

(1) 地震震级

表示震源释放能量的量度。

(2) 地震烈度

显示地面及建筑物受到破坏的程度。见表 2-11

2.15.1 Earthquakes

Basic Concepts

(1) Earthquake magnitude

Magnitude shows the energy released by the seismic focus.

(2) Earthquake (seismic) intensity

Seismic intensity shows the damaged situations of ground and buildings by the earthquake. See table 2-11.

表 2-11 不同地震烈度的危害

地震烈度	地面及建筑物受破坏的程度
1°~2°	人们一般感觉不到,只有地震仪才能记录到
3°	室内少数人能感到轻微的振动
4°~5°	人们有不同程度的感觉,室内物体有些摆动和有尘土掉落现象
6°	较老的建筑多要被损坏,个别有倒塌的可能;有时在潮湿疏松的地面上,有细小裂缝出现,少数山区发生土石散落
7°	家具倾覆破坏,水池中产生波浪;对坚固的住宅建筑有轻微的损坏,如墙上产生轻微的裂缝,抹灰层大片的脱落,瓦从屋顶掉下等;工厂的烟囱上部倒下;严重的破坏陈旧的建筑物和简易建筑物;有时有喷沙、冒水现象
8°	树干摇动很大,甚至折断;大部分建筑遭到破坏;坚固的建筑物墙上产生很大裂缝而遭到严重地损坏;工厂的烟囱和水塔倒塌
9°	一般建筑物倒塌或部分倒塌;坚固的建筑物受到严重破坏,其中大多数变得不适于使用;地面出现裂缝,山区有滑坡现象
10°	建筑严重毁坏;地面裂缝很多,湖泊、水库有大浪出现;部分铁轨弯曲变形
11°~12°	建筑普遍倒塌,地面变形严重,造成巨大的自然灾害

Table 2-11 DAMAGES OF DIFFERRENT SEISMIC INTENSITY

Seismic Intensity	Damaging Situation of Ground and Buildings
1°~2°	Earthquakemeter can record. Generally people no feeling
3°	Indoor a few people can feel a light vibration
4°~5°	People have some different feeling. Indoor somethings swing, dusts fall down
6°	More old buildings damaged, one or two might be collapsed. On wet-soft ground, small cracks appear. In a few mountain area soil and stones slided down
7°	Furnitures damaged and overturned. Pools waved. Light damages in strong dwelling buildings, such as small cracks in walls, render fallen off, tiles fallen down from roofing. Upper parts of chimney stalk collapsed. Old buildings and simply constructed buildings seriously damaged. Some times sand jetted and water oozed.
8°	Trees swing wide even broken off. Most buildings damaged. Big cracks appeared in strong building walls. Chimneies and water touers in factories collapsed

Seismic Intensity	Damaging Situation of Ground and Buildings
9°	Genenral buildings wholly or partly collapsed. Strong buildings seriously damaged, most of them not available. Ground cracked. In mountain area slides appeared
10°	Buildings seriously damaged. Ground most cracked. Lakes and reservoirs appear big waves. Some parts of Railway bended
11°~12°	All buildings collapsed. Ground seriously defomed. An extremely natural disaster happened

(3)基本烈度 该地区今后一定时期一般场地条件可能遭遇的最大地震烈度。

(4)设计烈度 设计采用的烈度

非常重要的建筑;关键性建筑如供水厂、电站、电信、医院、交通与消火中心等。设计烈度＞基本烈工 1°;

一般民用与工业建筑,设计烈度＝基本烈度;不重要的建筑,设计烈度可低于基本烈度 1°。

地震震级与震中的烈度关系见表2-12。

(3)Basic intensity In the area in future might happen the strongest intensity under the conditions of general site.

(4)Design intensity It's used in design.

Very important buildings; Key buildings: such as water supply, power station, communication, hospital, traffic and fire center, etc.: Design intensity > basic intensity, 1°;

General civil and industrial buildings: Design intensity = basic intensity; Not important buildings: Design intensity might be lower than basic intensity, 1°.

· The relation between magnitude and epicentre (震中) intensity see table 2-12.

表 2-12 震级及震中烈度的关系

The Relation between Magnitude and Epicentre Intensity

Magnitude 震级	1~2	3	4	5	6	7	8	8以上
Epicentre Intensity 震中烈度	1°~2°	3°	4°~5°	6°~7°	7°~8°	9°~10°	11°	12°

Note: The earthquake focus is underground 10~30km.
注:震源位于地下 10~30km。

(5)抗震构造

1)防震距离 例如 7°区两建筑间距不得小于 $1.2H$, H 为高建筑檐口距地面高度;

2)抗震缝 图 2-73,不同结构体部之间;不同高度(高差 6m 或以上)体部之间;不同刚度体部之间应设抗震缝。

变形缝细部方案见图2-74

缝宽 $B=\dfrac{H}{120}$(m), H 是低房高度。

(5)Aseismic Construction

1) Aseismic distance, between two buildings such as in 7° area, should not less than $1.2H$. H is the higher eaves height from ground.

2) Aseismic joint (seismic joint). Fig.2-73. Between the two parts of different structure; different height (one's height is higher than the other, 6m or more); different rigid.

Deformation joints' details see Fig.2-74

Widths of the seismie joints $B=\dfrac{H}{120}$ (m), H is the height of the lower building.

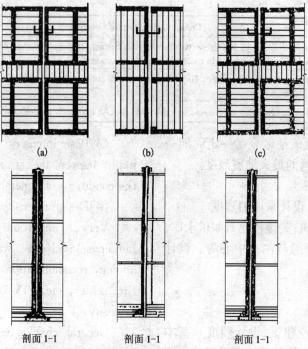

剖面 1-1　　　剖面 1-1　　　剖面 1-1

Fig. 2-73　Deformation joint schemata
图 2-73　变形缝方案
(a)双墙,基础不分开(适用于伸缩缝或称膨胀缝或称温度缝及房屋平面不复杂的抗震缝);
(b)墙与框架间变形缝,基础分开;(c)双墙基础分开的变形缝;(b)、(c)适用于沉降缝和房屋平面复杂的抗震缝)
(a)Double wall, foundation not separated (suitable to expansion joint, and aseismic joint of simple plane buildings);
(b)deformation joint between wall and frame, foundations separated; (c)Deformation joint between two foundations seprarated
(b),(c)suitable to settlement joint and aseismic joint of complicated plane buildings)

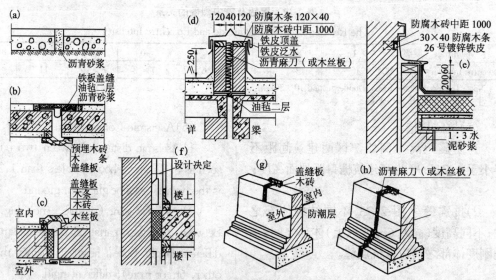

Fig. 2-74　Deformation joint details
图 2-74　变形缝细部
(a)底层地面;(b)楼板上下面;(c)外墙内外面;(d)平屋顶;(e)高低跨屋面;
(f)墙与楼板交接处;(g)基础不分开;(h)基础分开。
(a)ground floors;(b)Double sides of RC floors;(c)Double sides of outer walls;(d)flat roofs;(e)joint between high wall and low roof;
(f)joint between wall and floor;(g)separated walls on an un-separated footing;(h)separated footings and walls

3)构造柱与圈梁 图2-8(a)表示圈梁与构造柱组成了钢筋混凝土框架加强抗震与抗不均匀沉陷的刚度。

在钢筋混凝土框架结构或钢框架结构中无需构造柱和圈梁。

4)加强连接 在墙—墙、墙—楼板、楼板—楼板等构件之间用钢筋加强连接,图2-75示出了两例。

3) Constructional (tie) column and girth. Fig.2-8(a) shows girths and constructional columns combine to form a RC fram to strengthen the aseismic rigid and anti-unequal settlement rigid.

In RC fram construction or steel frame construction, there does not need any constructional columns and girths.

4) strengthening joints Between wall-wall, wall-floor and floor-floor, etc. using bars to strengthen their joints, Fig. 2-75 shows two examples.

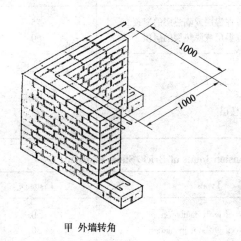
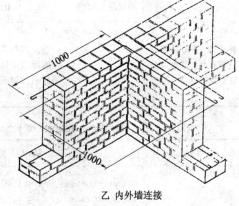

甲 外墙转角　　　　　　乙 内外墙连接

Fig.2-75　Bars strengthen joints

图 2-75　用钢筋加强连接

2.15.2 不均匀沉降

在湿陷或湿胀地区为避免不均匀沉降的危害,应设沉降缝(图2-73和图2-74)。

沉降缝宽度为30～70mm,在软土地区,宽度可以放宽到80～120mm甚至更多。

2.15.3 热变形

大气温度及太阳辐射的变化会引起建筑物热变形。

2.15.2 Unequal settlement

In wet-settlement or wet-expansion area, settlement joints should be set to avoid the damages by unequal settlement. (Fig. 2-73 and Fig.2-74)

Settlement joint width is 30～70mm. In soft soil the width might be 80～120mm even more.

2.15.3 Thermal deformation

The changes of air temperatures and solar radiation cause thermal deformations in buildings.

设伸缩缝(或称膨胀缝)是避免热变形破坏的主要方法。

表 2-13、表 2-14 列出了伸缩缝的最大距离。

Expansion joint (dilation joint, dilatation joint) is the main method to avoid the damage caused by the thermal deformation.

Table 2-13、Table 2-14 shows the max. distance of expansion joints.

表 2-13　砖石墙体伸缩缝的最大间距(m)

砌体类别	屋顶或楼层类别		间距
各种砌体	整体式或装配整体式钢筋混凝土结构	有保温层或隔热层的屋顶、楼层	50
		无保温层或隔热层的屋顶	30
	装配式无檩体系钢筋混凝土结构	有保温层或隔热层的屋顶、楼层	60
		无保温层或隔热层的屋顶	40
	装配式有檩体系钢筋混凝土结构	有保温层或隔热层的屋顶	75
		无保温层或隔热层的屋顶	60
普通黏土砖、空心砖砌体石砌体硅酸盐砖、硅酸盐砌块和混凝土砌块砌体	黏土或石棉水泥瓦屋顶		150
	木屋顶或楼层		100
	砖石屋顶或楼层		75

Table 2-13　The Max. Distance of Expansion Joints of Brick/Stone Walls(m)

Masonry Types	Roof or Floor Types		Distance
Various Masonries	Cast-in-site or Precast RC Construction	Roof with insulation	50
		Roof without insulation	30
	Precast RC Construction without Purlins	Roof with insulation	60
		Roof without insulation	40
	Precast RC Construction with Purlins	Roof with insulation	75
		Roof without insulation	60
Masonries of Clay Brick, Hollow Brick, Stone, Silicate Brick, Silicate block and Concrete Block	Clay tile or asbestos tile roof		150
	Wood roof or floors		100
	Brick or stone roof or floors		75

表 2-14 钢筋混凝土结构伸缩缝最大间距(mm)

结 构 类 别	室 内 或 土 中	露 天
钢筋混凝土整体式框架建筑	55	35
钢筋混凝土装配式框架建筑	75	50
装配式大型板材建筑	75	50

Table 2-14 The Max. Distance of Expansion Joint of RC Construction(mm)

Construction Types	Indoor or Underground	Open
Cast-in-site RC Frame Construction	55	35
Precast RC Frame Construction	75	50
Precast Panel Building	75	50

2.16 表面处理

2.16.1 万事万物均需表面处理

动物的皮毛脱(蜕)换,求偶展美;植物绿叶秋黄(红)、冬脱落、春萌芽、来夏再盛绿;人类的洗、沐、护肤、穿衣,化妆;建筑的装饰及环境美化等无不是表面处理。

为什么?为了可持续发展。

2.16.2 建筑及其环境表面处理应考虑的因素

表面处理必须有利于改善热、声、光环境;用电安全;防止电磁辐射;清除感应和摩擦静电的危害;防水及阻止湿气渗透;防火(慎用撞击起火花材料);严禁使用致癌性、过敏性及窒息性等有毒材料;防腐、防白蚁侵

2.16 Surface Treatments

2.16.1 All things have to take surface treated

Animals' regeneracy of feather, colorful showing for lovemaking; plants' green leaves faded in yellow or red in autumn, fallen in winter, germinating in spring and blooming again next summer; human washing, bathing, skin protection, dressing and prinking; building decoration and environment beautifying all are surface treatments.

Why? For sustainable development.

2.16.2 Factors should be considered for the surface treatments of buildings and environments

Surface treatments have to benefit improving the environments of heat, sound and light; electricity-use safety; forfending electromagnatic radiation; cancelling electrostatic damages of faradic or rubbed electricity; waterproofing and

害。表面终饰应耐久。

表面处理最重要因素之一就是要使全感官有最佳的美感。

2.16.3 表面处理法

(1)墙面

①清水墙勾缝,图2-76

stopping moisture penetration; fireproofing (impacting sparks of some materials which should be careful to use); poisonous materials such as causing carcinogenesis, allergic reaction and suffocation, etc. which should be strictly prohibited; forfending erosion and termites. Surface finishes should be durable.

To make all senses' perfect perception is one of the most important factors in surface treatments.

2.16.3 Surface treatment methods

(1) Wall surfaces

① Plain wall joint pointing, Fig. 2-76.

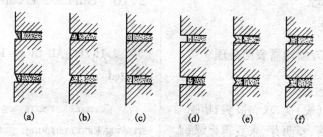

Fig. 2-76 Examples of plain wall joint pointing

图 2-76 清水墙勾缝例

(a)倒斜缝;(b)斜水缝;(c)平缝;(d)平凹缝;(e)V形缝;(f)弧凹缝

勾缝砂浆为水泥砂浆,体积比为1:1。

②水刷石;③瓷砖饰面;④混凝土块墙面(莫斯科大学);⑤彩色粉刷;⑥陶砖饰面;⑦油漆或喷漆;⑧雕塑;⑨花岗岩块贴面;⑩大理石块贴面;⑪无毒彩色塑料布贴面;⑫钢丝网抹灰;⑬斩假石;⑭镜面玻璃幕墙;⑮金属涂层玻璃幕墙;⑯土蛋饰面;⑰挂轴;⑱壁画如飞天;⑲活动风景画;⑳墙面绿化;㉑剪纸;㉒木饰面。

joint pointing mortar is cement-sand mortar, volume rate 1:1 cement:sand.

②Granitic plaster; ③Glazed tile finishing; ④Concrete block wall(Moscow University); ⑤Colour render; ⑥Ceramic tile finishing; ⑦Painting/spray painting; ⑧ Carving; ⑨ Granitic block finishing; ⑩ Marble block finishing; ⑪Non-poisonous color plastic film finishing; ⑫plaster on metal lath; ⑬Artificial stone; ⑭Mirror curtain wall; ⑮Metal coating glass curtain wall; ⑯ Soil ball finishing; ⑰ Hanging scroll; ⑱ Mural painting such as

(2)地层及楼层表面

①现浇混凝土地面；②现浇水磨石地面；③预制混凝土块地面；④预制水磨石地面；⑤马赛克(锦砖)铺面；⑥缸砖铺面；⑦黏土砖地面；⑧土地面；⑨木地面(多种图案)；⑩橡胶地面；⑪菱古土地面(菱古土浆加木屑(锯末)拌成；表面可油漆)；⑫耐酸地面,如硅酸钠(水玻璃)混凝土；沥青混凝土；沥青砂浆；⑬植草地面；⑭自然草地面；⑮花岗岩块地面；⑯大理石块地面；⑰铸铁板地面；⑱撞击不起火花地面：石灰石,白云石,沥青,塑料,橡胶,木材,铜,铝,铅；⑲斧砍面；⑳屋顶花园；㉑通风屋顶；㉒多功能屋顶；㉓吊顶；㉔普通顶棚如抹灰顶棚；㉕贴饰面；㉖喷饰面；㉗中国藻井天花；㉘吊饰。

(3)立体空间处理

①室内景观；②室外景观；③激光成像；④立体雕塑；⑤立体绿化。

例：图 2-77~2-84。

其它实例见相关彩图。

flying Apsaras (Flying fairy); ⑲ Moving landscape pictures; ⑳ Wall greening; ㉑ Paper cut; ㉒ Wood finishing.

(2) Ground and Floor Surfaces

① Cast - in - site concrete ground; ②Cast-in- place terrazzo floor; ③ Precast concrete block floor; ④Precast terrazzo floor; ⑤ Mosak pavement; ⑥ Clink (tile) floor, quarry tile floor; ⑦ Clay brick floor; ⑧ Soil floor; ⑨ Wooden floor (various patterns); ⑩ Rubber floor; ⑪ Magneste flooring; ⑫ Acidproof floor (acid-resisting floor). such as sodium silicate (water-glass) concrete; Asphelt concrete; Asphelt mortar; ⑬Grassing ground; ⑭ Natural grass ground; ⑮ Granitic block ground; ⑯Marble block ground; ⑰Casted iron block ground; ⑱ Sparkless floor under impacting: lime stone , dolomite, asphelt, plastic, rubber, wood, copper, aluminium, lead are sparkless materials under impacting (striking); ⑲Axed surface; ⑳Roof garden; ㉑ Ventilation roof; ㉒ Multi - function roof; ㉓ Hanged ceiling; ㉔ Common ceilling such as plastered ceiling; ㉕Glued finishing; ㉖sprayed finishing; ㉗ Chinese sunk panel; ㉘ Hanging decoration.

(3) Stereoscopic Space Treatments

①Indoor landscape; ②Outdoor landscape; ③ Laser picturing; ④Stereoscopic carvings; ⑤Three-dimensional greening.

Examples：

Fig.2-77~2-84.

Other practical examples see relative color pictures.

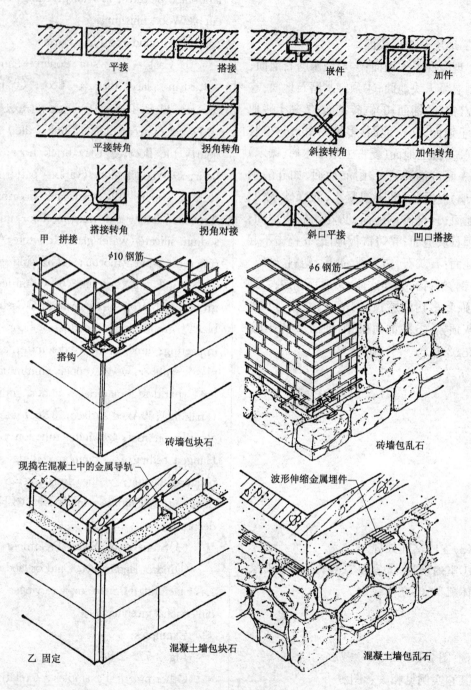

Fig. 2-77 Stone block finish
图 2-77 块石贴墙面

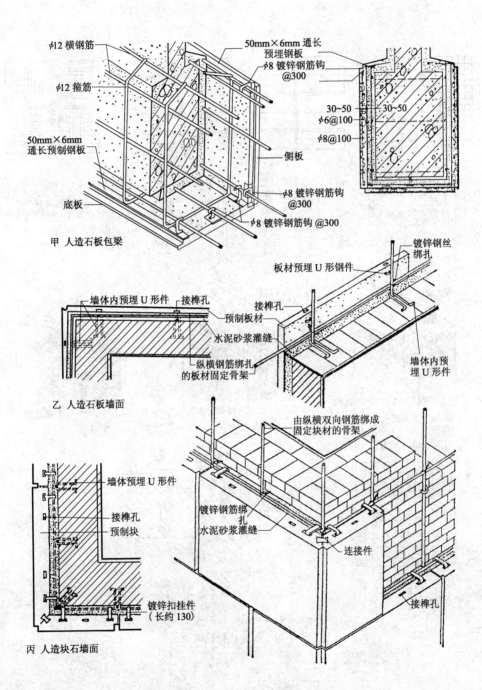

Fig. 2-78　Artificial stone finish
图 2-78　人造石贴墙面

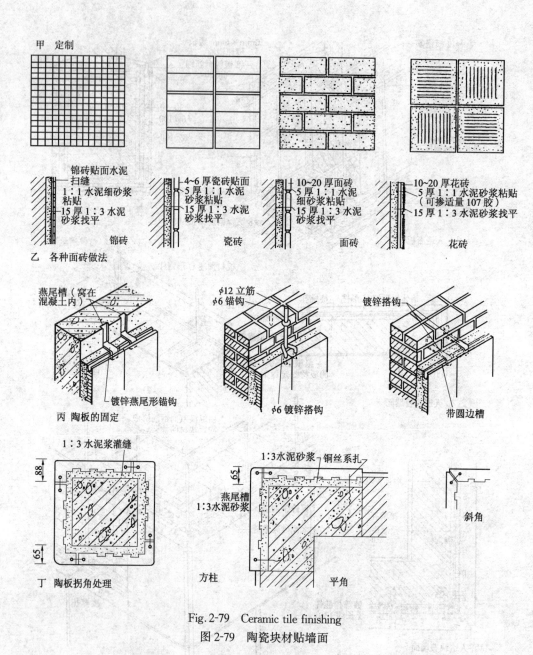

Fig. 2-79 Ceramic tile finishing
图 2-79 陶瓷块材贴墙面

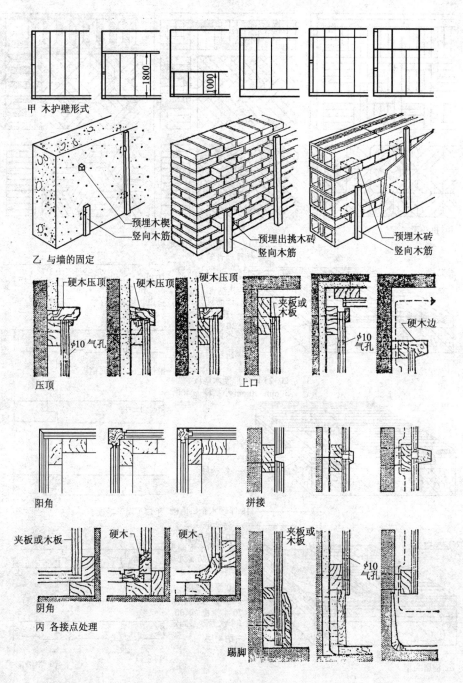

Fig. 2-80　Wood dado
图 2-80　木墙裙

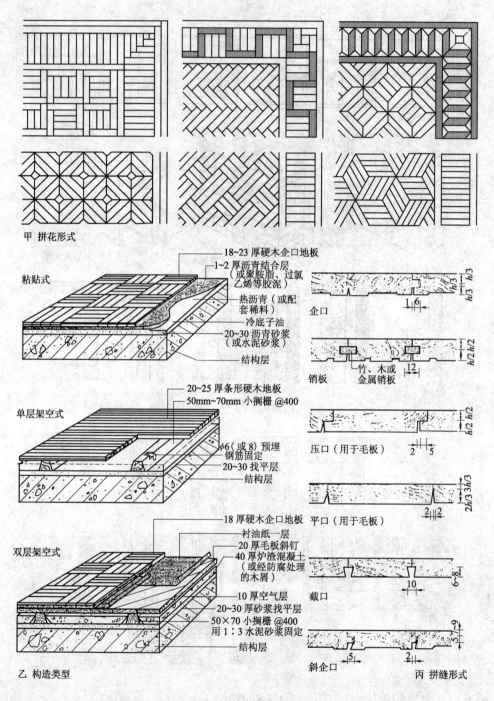

Fig. 2-81 Pattern floor by wood block
图 2-81 硬木拼花地板

Fig. 2-82 Cast-in-place terrazzo floor under doing
图 2-82 现浇水磨石地面在施工中

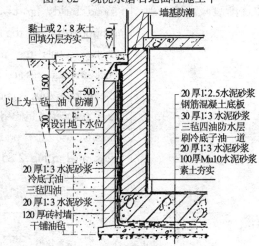

Fig. 2-83 Waterproof treatment of basement wall
图 2-83 地下室墙面防水处理

Fig. 2-84 Outer plastering of model sliding building
图 2-84 滑模建筑外抹灰

2.17 厅堂音质

2.17.1 室内声音的表演

(1)声形成与衰减

当一恒定声源在室内传播声音时,其产生的连续声波会向各个方向散播。声波撞击各种表面,部分被吸收,部分被反射。声源传出声能的速率与室内周围及所容纳的表面吸收声的速率相等时,该声的平均强度达到最大值并持续下去。若将声源切断,该室内声不会立即停止而是被吸收逐渐衰减。这一现象示于图 2-85。

2.17 Acoustical Properties of Halls

2.17.1 Sound playing in a room

(1)Build-up and decay of sound

When a constant source of sound is introduced into a room, each successive wave it produces spreads in all directions. As it strikes the various surfaces in the room, it is partially absorbed and partially reflected from the several surfaces it strikes. The average intensity of the sound builds up to a maximum at which it continues because the rate of introduction of sound energy equals the rate of absorption by the enclosing surfaces and the contents of the room. If the sound source is cut off, the sound in the room does not cease immediately but gradually decays because of absorption. This phenomenon is illustrated in Fig. 2-85.

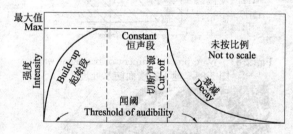

Fig. 2-85 Build-up, continuation and decay of sound in a room
图 2-85 室内声的起始、延续与衰减

(2)回声

当反射声可清楚地被听出对原声的重复,此反射声即为回声。这两种声的时间差至少在 1/20 秒(50 毫秒)以上才能形成回声。该时间差小于此数。则反射声仅对原声起加强作用,故小房间不会产生回声。

(2)Echoes

When the reflection of a sound is heard as a distinct repetition of the original sound, the reflection is called an echo. For an echo to be formed, the time difference between the two sounds must be at least 1/20 of a second (50 ms). If the difference is smaller than this the reflected sound merely reinforces the original

(3)混响声

混响声由来自墙、地面和天棚的反射声形成。演说用房间混响声应尽快消失以免干扰清晰度。音乐演奏用房间要有一个合适的混响音质范围。图 2-86 为室内直达声与混响声的合成曲线。

(3) Reverberant Sound

Reverberant sound is formed by the reflections from the walls, floor and ceiling. A space designed for speech should ensure that the reverberant sound dies away rapidly and not confuses the definition. A space designed for music may have a range of appropriate reverberant qualities. Fig.2-86 shows the sum curve of direct sound and reverberant sound inside a room.

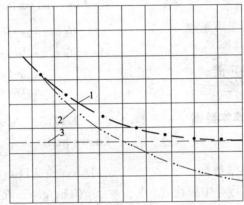

Fig.2-86　The sum curve of direct sound and reverberant sound inside a room

图 2-86　室内直接声与混响声的叠加(合成)曲线

1.合成曲线　2.直接声级线　3.混响声级线

1.Sum　2.Direct sound level　3.Reverberant sound level

2.17.2　厅堂音质

(1)亲切　直达声与 30 毫秒以内的反射声合成在一起会引起对原声的一种亲切感的幻觉。

听众席做成后排高于前排的斜度；观众厅前面的墙面做反射面以及合适的天棚形式都是改善亲切感的有效措施。

(2)温暖或丰满度　温暖或丰满度与混响声含量有关特别是低频成份。亲切感太多会降低清晰度，反之亦然。

(3)混响时间　(RT)即声音从初始声级

2.17.2　Acoustical properties of halls

(1) Intimacy　Direct sound taken together with the reflections within 30 ms (milli second) which may cause an intimate illusion to the original sound.

Raking the seating of the audience, reflecting surfaces near the front and an appropriately shaped ceiling all are available to improve intimacy.

(2) Warmth or Fullness　This relates to the proportion of reverberant sound especially the low frequency content. Too much warmth can reduce definition, and vice versa.

(3) ReverberationTime(RT)　RT is the time

衰减60 dB(衰减到原声级的百万分之一)所经历的时间(秒)。有名的混响时间估算式——Sabine公式:

$$RT = \frac{0.16V}{A} \quad (秒)$$

RT:混响时间(秒);
V:厅堂容积(m^3);
A:总吸声量(m^2 或 sabin);
$A = S_1\alpha_1 + S_2\alpha_2 + S_3\alpha_3 + \cdots + Sn\alpha_n$
$S_1, S_2, S_3\cdots$室内各表面面积(m^2);

$\alpha_1, \alpha_2, \alpha_3\cdots$各表面吸声系数

$$\alpha = \frac{未反射的声能}{入射声能}$$

2.17.3 著名观演厅声学数据

(1)某些著名观演厅声学数据
〔所给混响时间(RT)乃对中频而言,并假设观众满座〕

名称	体积(m^3)	听众容量	每位听众占有的容积(m^3)	混响时间(s)
1. 音乐厅				
阿姆斯特丹音乐厅(1887)	18 700	2 200	8.5	2.0
柏林交响乐团音乐厅(1963)	26 000	2 200	11.8	2.0
波斯顿交响乐厅(1900)	18 500	2 600	7.1	1.8
来比锡牛斯格旺德荷观演厅(1886)	10 600	1 560	6.8	1.55
伦敦皇家音乐节音乐厅(1951)	22 000	3 000	7.3	1.47

in second for a sound to decay by 60dB (i.e. one millionth of its original loudness). The now well known Sabine formula for evaluating RT is given as:

$$RT = \frac{0.16V}{A} \quad (\text{second})$$

RT: Reverberation time in second(s);
V: Volume of the hall (m^3);
A: total absorption (m^2 or sabin);
$A = S_1\alpha_1 + S_2\alpha_2 + S_3\alpha_3 + \cdots + Sn\alpha_n$
$S_1, S_2, S_3\cdots$ The composed areas of the total internal surface of the hall(m^2);

$\alpha_1, \alpha_2, \alpha_3 \cdots$ These areas possess absorption coefficients

$$\alpha = \frac{\text{sound energy not reflected}}{\text{sound energy incident}}$$

2.17.3 Acoustical data of some well known halls

(1) Acoustical Data of Some Well Know Halls
(The given RT is for middle frequency and assuming capacity audience)

Name	Volume (m^3)	Audience capacity	Vol/aud. seat (m^3)	RT (s)
1. CONCERT HALLS				
Amsterdam, Concertgebouw (1887)	18 700	2 200	8.5	2.0
Berlin, Philharmonic Hall(1963)	26 000	2 200	11.8	2.0
Boston, Symphony Hall (1900)	18 500	2 600	7.1	1.8
Leipzig, Neues Gewandhaus (1886)	10 600	1 560	6.8	1.55
London, Royal Festival Hall (1951)	22 000	3 000	7.3	1.47

续表

名称	体积 (m³)	听众容量	每位听众占有的容积(m³)	混响时间(s)
1. 音乐厅				
纽约开纳奇音乐厅(1891)	24 200	2 760	8.8	1.7
纽约交响乐团音乐厅(1962)	24 430	2 640	9.3	2.0
昆士兰文化中心音乐厅	22 000	2 000	11.0	2.4
悉尼歌剧院音乐厅(1973)	25 860	2 700	9.6	2.1
悉尼市政厅(1890)	2300	2000	11.5	2.2
维也纳格罗色尔音乐厅(1870)	15 000	1 680	8.9	2.05
2. 歌剧院				
伦敦皇家歌剧院(1858)	12 240	2 210	5.5	1.1
米兰歌剧院(1778)	11 250	2 490	4.5	1.2
纽约大都市歌剧院(1883)	19 500	3 780	5.2	1.2
巴黎国家歌剧院(1875)	9 960	2 230	4.5	1.1

续表

Name	Volume (m³)	Audience capacity	Vol/aud. seat (m³)	RT (s)
1. CONCERT HALLS				
New York, Carnegie Hall (1891)	24 200	2 760	8.8	1.7
New York, Philharmonic Hall(1962)	24 430	2 640	9.3	2.0
Queensland, Cultural Center Concert Hall	22 000	2 000	11.0	2.4
Sydney, Opera House Concert Hall(1973)	25 860	2 700	9.6	2.1
Sydney, Town Hall(1890)	2300	2000	11.5	2.2
Vienna, Grosser Musikvereinssaal (1870)	15 000	1 680	8.9	2.05
2. OPERA HOUSES				
London, Royal Opera House (1858)	12 240	2 210	5.5	1.1
Milan, Teatro alla Scala (La Scala)(1778)	11 250	2 490	4.5	1.2
New York, Metropolitan Opera House (1883)	19 500	3 780	5.2	1.2
Paris, Theatre National de L'Opera(1875)	9 960	2 230	4.5	1.1

续表

名称	体积 (m³)	听众容量	每位听众占有的容积(m³)	混响时间(s)
2. 歌剧院				
维也纳歌剧院 (1957)	10 660	1 940	5.5	1.3

注：以上各厅中，许多厅允许有站立的听众，上表听众容量一栏已经根据正式座位进行了修整。

(2) 不同表面吸声系数(α)（见表2-15）(1)、(2)

表 2-15(1) 不同表面吸声系数(α)

表面		频率		
		125	500	2000
混凝土	结构混凝土的光平面 砌块光平面 水磨石光平面	0.01	0.02	0.02
砖砌体	磨光砖面	0.03	0.03	0.05
混凝土块材		0.10	0.06	0.09
纤维板	厚13mm,有坚固基层	0.05	0.15	0.30
	同前具有空气层	0.30	0.30	0.30
石膏板	搁栅支承,有空气层	0.30	0.05	0.07
玻璃	大型平板玻璃	0.20	0.40	0.02
	玻璃窗	0.35	0.18	0.07
胶合板	夹空气层	0.30	0.15	0.10
帘幕	轻型与墙贴靠	0.03	0.10	0.30

续表

Name	Volume (m³)	Audience capacity	Vol/aud. seat (m³)	RT (s)
2. OPERA HOUSES				
Vienna, Staatsoper (1957)	10 660	1 940	5.5	1.3

Note: In the above halls many of them allow audience to stand, the above data have been corrected according to the normal seats.

(2) ABSORPTION COEFFICIENTS(α) OF VARIOUS SURFACES (Table 2-15)(1)、(2)

Table 2-15(1) Absorption Coefficients(α) Various Surfaces

SURFACE		FREQUENCY (Hz)		
		125	500	2000
Concrete	structural smooth stonework smooth terrazzo smooth	0.01	0.02	0.02
Brickwork	glazed	0.03	0.03	0.05
Concrete block		0.10	0.06	0.09
Fibre-board	13mm - solid backing	0.05	0.15	0.30
	as above with airspace	0.30	0.30	0.30
Gypsum	joists and panels airspace	0.30	0.05	0.07
Glass	heavy plate	0.20	0.40	0.02
	window	0.35	0.18	0.07
Plywood panels	on airspace	0.30	0.15	0.10
Curtains	light, contacting wall	0.03	0.10	0.30

表 2-15(2) 材料吸声系数 α 及吸声单位(m^2, sabin)

材料及其安装情况	吸声系数 α					
	125Hz	250Hz	500Hz	1000Hz	2000Hz	4000Hz
清水砖墙	0.05	0.04	0.02	0.04	0.05	0.05
砖墙上抹灰(光面)	0.024	0.027	0.03	0.037	0.036	0.034
抹灰拉毛,面涂漆	0.04	0.04	0.07	0.024	0.09	0.05
木板墙(紧贴实墙)	0.05	0.06	0.06	0.10	0.10	0.10
纤维板厚 1.25cm(紧贴实墙)	0.05	0.10	0.15	0.25	0.30	0.30
同上,表面涂漆	0.05	0.10	0.10	0.10	0.10	0.15
三夹板后空气层为5cm,龙骨间距 50cm×50cm	0.206	0.737	0.214	0.104	0.082	0.117
同上,空气层中填矿棉($8kg/m^2$)	0.367	0.571	0.279	0.118	0.093	0.116
三夹板后空气层为10cm,龙骨间距为 50cm×45cm	0.597	0.2362	0.181	0.05	0.041	0.082
五夹板后空气层为10cm,龙骨间距 50cm×45cm,涂三道油	0.199	0.10	0.25	0.057	0.062	0.191
三夹板穿孔(ϕ5mm)孔距4cm,后空气层为10cm,板背后贴一层龙头细布,板后填矿棉($8kg/m^2$)	0.673	0.731	0.507	0.287	0.191	0.166
木丝板(厚3cm)后空10cm,龙骨间距 45cm×45cm	0.09	0.36	0.62	0.53	0.71	0.87
同上,后空 5cm	0.05	0.30	0.81	0.63	0.70	0.91
聚氨脂泡沫塑料,厚2cm	0.055	0.067	0.16	0.51	0.84	0.65
同上,厚4cm	0.12	0.22	0.57	0.77	0.77	0.76
丝绒幕($0.65kg/m^2$),离墙10cm	0.06	0.27	0.44	0.50	0.40	0.35
同上,离墙20cm悬挂	0.08	0.29	0.44	0.50	0.40	0.35
玻璃(紧贴实墙)	0.01		0.01		0.02	
玻璃窗扇(125cm×35cm),玻璃厚3mm	0.35	0.25	0.18	0.12	0.07	0.04
同上,玻璃厚6mm	0.01		0.04		0.02	
通风口及类似物、舞台开口	0.16	0.20	0.30	0.35	0.29	0.31
普通抹灰吊顶(上有大空间)	0.20		0.10		0.04	
钢丝网抹灰吊顶(厚5cm)	0.08	0.06	0.05	0.04	0.04	0.04
加玻璃纤维筋的塑料反射板(厚1.5mm,吊在空中)	0.45	0.23	0.10	0.37	0.37	0.37
光面混凝土(厚10cm以上)	0.01	0.01	0.02	0.02	0.02	0.03
木地板(有龙骨架空)	0.15	0.11	0.10	0.07	0.06	0.07
毛地毯厚1.1cm,铺在混凝土上	0.12	0.10	0.28	0.42	0.21	0.33
橡皮地毯厚5mm,铺在混凝土上	0.04	0.04	0.08	0.12	0.03	0.10
听众席(包括听众、乐队所占地面,加周边宽一米的走道)	0.52	0.68	0.85	0.97	0.93	0.85
空听众席(条件同上,座椅为软垫的)	0.44	0.60	0.77	0.89	0.82	0.70
听众(坐在软垫椅上,按每人计算)	0.19	0.40	0.47	0.47	0.51	0.47
软垫座椅(每个)	0.12		0.28		0.32	0.37
乐队队员带着乐器(坐在椅子上,每人)	0.38	0.79	1.07	1.30	1.21	1.12
听众(坐在硬垫椅上,每人)	0.27	0.21	0.37	0.46	0.54	0.46
木板硬座椅(每人)	0.07	0.03	0.08	0.10	0.08	0.11

Table 2-15(2) Material Absorption Coefficient α and Sound Absorption Unit(m², sabin)

Materials and Construction	Absorption Coefficient α					
	125Hz	250Hz	500Hz	1000Hz	2000Hz	4000Hz
Brick plain wall	0.05	0.04	0.02	0.04	0.05	0.05
Brick wall rendered (smooth)	0.024	0.027	0.03	0.037	0.036	0.034
Stucco, surface painted	0.04	0.04	0.07	0.024	0.09	0.05
Wood wainscoting (closely veneered to the solid wall)	0.05	0.06	0.06	0.10	0.10	0.10
Fibre board, 1.25cm t(Ditto)	0.05	0.10	0.15	0.25	0.30	0.30
Ditto, surface painted	0.05	0.10	0.10	0.10	0.10	0.15
5cm air gap behind three ply board, joist pitch 50cm×50cm	0.206	0.737	0.214	0.104	0.082	0.117
Ditto, air gap filled by mineral wool(8kg/m²)	0.367	0.571	0.279	0.118	0.093	0.116
10cm air gap behind three-ply board, joist pitch 50cm×45cm	0.597	0.2362	0.181	0.05	0.041	0.082
10cm air gap behind five-ply board, joist pitch 50cm×45cm, 3-layer painting	0.199	0.10	0.25	0.057	0.062	0.191
10cm air gap filled by mineral wool(8kg/m²) behind three-ply perforated panel (φ5mm), hole pitch 4cm, a layer fine cloth attached to the panel back	0.673	0.731	0.507	0.287	0.191	0.166
Wood-fibre board, 3cm t, 10cm air gap behind the board, joist pitch 45cm×45cm	0.09	0.36	0.62	0.53	0.71	0.87
Ditto, air gap 5cm	0.05	0.30	0.81	0.63	0.70	0.91
Polyurethane plastic foamed, 2cm t	0.055	0.067	0.16	0.51	0.84	0.65
Ditto, 4cm t	0.12	0.22	0.57	0.77	0.77	0.76
Velvet(velour) curtain(0.65kg/m²), 10cm before the wall	0.06	0.27	0.44	0.50	0.40	0.35
Ditto, 20cm before the wall	0.08	0.29	0.44	0.50	0.40	0.35
Glass (closely veneered to the solid wall)	0.01		0.01		0.02	
Glazing casement(125cm×35cm), glass 3mm t	0.35	0.25	0.18	0.12	0.07	0.04
Ditto glass 6mm t	0.01		0.04		0.02	
Vent & the likes, stage open	0.16	0.20	0.30	0.35	0.29	0.31
A large air space behding a suspended ceiling plastered	0.20		0.10		0.04	
5cm air gap behind plaster on metal lath	0.08	0.06	0.05	0.04	0.04	0.04
Plastic reflection panel of glass fibre reinforced(1.5mm t, hung in space)	0.45	0.23	0.10	0.37	0.37	0.37
Smooth concrete(t>10cm)	0.01	0.01	0.02	0.02	0.02	0.03
Wood floor(on joist frame)	0.15	0.11	0.10	0.07	0.06	0.07
Wool carpet(1.1cm t), on concrete	0.12	0.10	0.28	0.42	0.21	0.33
Rubber carpet(5mm t), on concrete	0.04	0.04	0.08	0.12	0.03	0.10
Auditory area (incl the area occupied by auditors, orchestra pit & the 1m w perimeter walkway)	0.52	0.68	0.85	0.97	0.93	0.85
Auditory empty(Ditto, soft chairs)	0.44	0.60	0.77	0.89	0.82	0.70
Auditors(in soft chair, per capta)	0.19	0.40	0.47	0.47	0.51	0.47
Each soft chair	0.12		0.28		0.32	0.37
Orchestra member with instrument (in chair, per capita)	0.38	0.79	1.07	1.30	1.21	1.12
	0.27	0.21	0.37	0.46	0.54	0.46
Wood hard chair(each one)	0.07	0.03	0.08	0.10	0.08	0.11

2.17.4 吸声

吸声是改善厅堂音质尤其是混响时间的主要方法之一。

吸声就是将声波机械能转换成热能。常用吸声器有三种：多孔吸声器，板片——空气层吸声器和亥姆霍兹吸声器。

(1) 多孔吸声器

玻璃棉、地毯、帘子、衣服、任何有连孔网络的物件都能通过摩擦将声能转化为热能。

多孔吸声器对高频吸声效率比对低频吸声效率好。

这种吸声器背后设一空气层，可稍改善低频段吸声。轻质材料如泡沫树脂，其内部小孔互不相通，一般吸声效率都不高。

多孔吸声器若被油漆涂层覆盖吸声效率将大为降低。

(2) 板片—空气层吸声器

板片—空气层吸声器由一个非多孔板材，并且板后留一空气层安装在固体基底上组成。空气层相当于一层弹性层。这种吸声器在63~250Hz范围吸声好。

用多孔材如玻璃丝棉填入空气层，可使吸声范围增到500Hz。

(3) 亥姆霍兹或空腔共振吸声器

这是最广泛的一种穿孔板后夹一空气层安装在实体构件上的空腔共振吸声器。其共

2.17.4 Sound absorption

Sound absorption is one of the main methods of improving the hall acoustic properties especially the reverberation time.

The aim of all acoustical absorbers is to convert mechanical energy in the sound wave into heat. There are three types of absorbers i.e., Porous, Panel-air space and Helmholtz absorbers which are usually used.

(1) Porous Absorbers

Glasswool, carpets, curtains and people's clothing, in fact anything with a network of interlocking pores can convert sound energy into heat by friction (rubbing).

The absorption of porous absorber is more efficient at high frequencies than at low frequencies.

The absorption can be slightly increased into low frequencies by mounting them with an air space behind. Light weight materials like foamed resins (泡沫树脂), whose pores do not interconnect, do not in general have high absorption.

If a coat of paint is applied to a porous absorber, the absorption of sound is considerably reduced.

(2) Panel-air space Absorber

The panel-air space absorber comprises a layer of non-porous material mounted with an air space between it and a solid backing. The air space is acting as a spring. This system absorbs well in the low frequencies region 63 to 250Hz.

The inclusion of porous absorbers such as glasswool in the space behind the panel can extend the range of absorption up to 500 Hz.

(3) Hemholtz or Cavity Resonators

This is the most widespread cavity resonator, it uses a perforated panel mounted

振频率可由板厚、孔径、孔距及空气层厚度控制。

这类吸声器有很强的频率专定性，可有效地吸收有干扰性的最频波而对其余的声频谱影响又不大。

图 2-87(a)、(b)、(c)分别为上述三种吸声器剖面例。

图 2-87(f)~(h)示出了上述三种吸声器的典型吸声特性，可以看出，当声频愈近其共振频率则被吸声能越多。吸声结构的共振频率最好是现场测量。对于最常用的亥姆霍兹吸声结构可按下述公式估算其共振频率(f_0)。

$$f_0 = \frac{V}{2\pi}\sqrt{\frac{P}{L(t+\delta)}} \quad (\text{Hz})$$

V:空气中声速：34000 cm/s；

δ:修正系数，设圆孔直径为 d，则 $\delta = 0.8d$；

P, L, t 含义见前文。（计算时，L, t 以 cm 为单位）

举例：设一亥姆霍兹共振吸声器，$L = 10\text{cm}$, $d = 0.8\text{cm}$, $B = 2\text{cm}$, $t = 0.4\text{cm}$，圆孔按正方形排列，求共振频率 f_0。

[解]：首先求穿孔率 $P = \frac{\pi}{4}\left(\frac{d}{B}\right)^2 = \frac{3.14}{4}\left(\frac{0.8}{2.0}\right)^2 = 0.125$

于是，$f_0 = \frac{V}{2\pi}\sqrt{\frac{P}{L(t+\delta)}} = \frac{34000}{2\times 3.14} \times \sqrt{\frac{0.125}{10\times(0.4+0.8\times 0.8)}} = 590 \text{ Hz}$

with an airspace between the panel and a solid backing. The resonant frequency of the cavity resonator can be controlled by the thickness of the panel, diameter of holes, spacing of holes and the size of air space behind the panel.

These resonators are very frequency specific, can be usefully used to absorb a particularly annoying room mode without greatly affecting the performance of the room throughout the rest of the sound spectrum.

Fig. 2-87(f)~(h) have shown (showed) the above three absorbers' typical absorptance characters, we can see when the sound is the more near the resonance frequency of the absorber the more sound energy absorbed. For finding the resonance frequency "f_0" of absorption construction, the optimal method is to measure on site. For the most used Helmheltz absorber the resonance freguency f_0 can be calculated by the following formula:

$$f_0 = \frac{V}{2\pi}\sqrt{\frac{P}{L(t+\delta)}} \quad (\text{Hz})$$

V = Sound velocity in air: 34000 cm/s;

δ = An correct factor, if "d" is the diameter of the round hole, the $\delta = 0.8d$;

P, L, t see the above text. (in calculation, L and t take "cm" as the unit)

An example: A Helmholtz absorption construction with $L = 10\text{cm}$, $d = 0.8\text{cm}$, $B = 2\text{cm}$, $t = 0.4\text{cm}$, the round holes are square arrangement, please find the f_0.

[Resolution]: First, $P = \frac{\pi}{4}\left(\frac{d}{B}\right)^2 = \frac{3.14}{4}\left(\frac{0.8}{2.0}\right)^2 = 0.125$

Then, $f_0 = \frac{V}{2\pi}\sqrt{\frac{P}{L(t+\delta)}} = \frac{34000}{2\times 3.14} \times \sqrt{\frac{0.125}{10\times(0.4+0.8\times 0.8)}} = 590 \text{ Hz}$

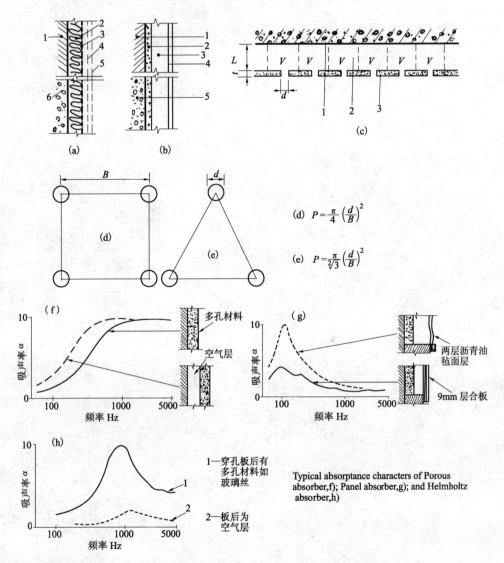

Fig 2.87 Section examples of the above absorbers and their typical absorption curves
图 2-87 上述吸声器剖面例及其吸声典型曲线

(a)多孔吸声器:1-砖墙;2-玻璃丝(2~5cm);3-玻璃布;4-金属或塑料窗纱;5-冲孔薄金属片;
6-混凝土墙(b)板材-空气层吸声器:2-抹灰;3-空气间层;4-板材(4~9mm);
(c)亥姆霍兹或空腔共振吸声器:1-RC楼板;2-空气层(5~20cm);3-穿孔板,t-板厚4~6mm;d-圆孔直径6~8mm;
L-空气层(空腔)厚5~20cm;B-孔距,一般取20mm;P-穿孔率,即穿孔面积比总面积,(d)、(e)为计算法。
(a)Porous Absorbers:1-brick wall; 2-glass fibre(2~5cm);3-glas-fibre cloth;
4-metal or plastic window screen;5-holed thin steel metal;6-concrete wall;
(b)Panel-air space Absorbers: 1-brick wall;2-plaster;3-air space(air gap 2~5cm);4-panel(4~9mm);
5-concrete wall (c) Hemholtz or Cavity Resonators:1-RC floor; 2-air space 5~20cm;
3-holed panel, t-thickness 4~6mm;d-round hole diameter d=6~8mm;L-air space,5~20cm;B-hole spacing,
general 20mm;P-holed rate i.e. holed area/total area. (d)&(e)are the calculation methods of p.

注:
1. 本书建筑构造图除部分为作者绘制外,多数引自原南京工学院唐厚炽教授等编著《建筑构造》通用教材;少数引自[美]《Building Construction》。
2. 书中有些简单英语单词没有配应汉语,目的是迫使学生查词典阅读。